D0583790

NON-EUCLIDEAN GEOMETRY

STEFAN KULCZYCKI

Translated from the Polish by
STANISLAW KNAPOWSKI

DOVER PUBLICATIONS, INC.
Mineola, New York

Bibliographical Note

This Dover edition, first published in 2008, is an unabridged republication of the work originally published as Volume 16 in the International Series of Monographs on Pure and Applied Mathematics by Pergamon Press, New York, in 1961.

Library of Congress Cataloging-in-Publication Data

Kulczycki, Stefan.
Non-Euclidean geometry / Stefan Kulczycki ; translated from the Polish by Stanislaw Knapowski.—Dover ed.
p. cm.
Prev ed. of: New York : Pergamon Press, 1961.
Includes index.
ISBN-13: 978-0-486-46264-6
ISBN-10: 0-486-46264-1
1. Geometry, Non-Euclidean. I. Title.

QA685.K8 2008
516.9—dc22

2007042750

Manufactured in the United States of America
Dover Publications, Inc., 31 East 2nd Street, Mineola, N.Y. 11501

CONTENTS

PREFACE

Non-Euclidean geometry is a young science; the date of its birth may be taken as being 1830, when Lobatchevsky's first publication appeared. This new doctrine, although at first treated with indifference, not to say ridicule, became, over the course of half a century, generally accepted among mathematicians. W. K. Clifford, the outstanding English scholar, refers to Lobatchevsky as the "Copernicus of geometry", thus manifesting his conviction of the immense importance of non-Euclidean concepts for science and for the formulation of a *Weltanschauung*: for the name of Copernicus is associated not only with his scientific discoveries, but above all with his transformation of our ideas about the universe. The conclusions of such penetrating thought could not long be left in the exclusive possession of scientists, and it is no wonder that they soon began to spread out among an ever-widening public, arousing general interest. Quite a number of books dealing with this subject appeared—this subject which entails so many peculiar difficulties but which may be understood by anybody familiar with the basic principles of elementary geometry.

My chief aim is here to give an accessible exposition of the principles of non-Euclidean geometry. By "accessible", I mean something not too remote from the formulations and arguments of elementary geometry. With this in mind I have made no mention of the connections with projective geometry, nor laid any stress on group concepts. Moreover, I have discarded topics related to continuity and treated them, if at all, rather objectively. This is precisely the standpoint of the creators of non-Euclidean geometry. It is my intention to give an overall picture of the whole subject, and in particular to prove the theorem on the angle of parallelism.

The proofs of this theorem are of two kinds:

(i) Stereometric and planimetric proofs based on the properties of horocycles, and

(ii) the so-called elementary proofs which make use only of the simplest properties of the plane.

These proofs, of which that of Straszewicz appears to be the perfect example, do not refer to any further theorems but use ingenious limiting processes and require the solution of functional equations. These I do not consider to be "accessible".

We have chosen the proof based on the study of the horosphere not only to mark the occasion of the centenary of Lobatchevsky's death but also in view of its advantages.

The present book contains three chapters. The knowledge of elementary geometry gained in schools and colleges will be sufficient for an understanding of the first two. The third chapter demands a certain familiarity with the principles of trigonometry; for §§ 28 and 29 elements of analytical geometry are also needed.

Chapter I gives some information about the history of geometry; chapters II and III contain a systematic exposition, without referring back to chapter I, of the principles of non-Euclidean geometry.

The exposition has been presented in such a way that the reader may limit himself to an examination of §§ 8-20 and obtain thereby a certain completeness of information. The remaining sections deal with non-Euclidean trigonometry.

STEFAN KULCZYCKI

Warsaw, 1956

CHAPTER I

FROM THE HISTORY OF GEOMETRY

§ 1. Earliest times

Geometry probably originated in Ancient Egypt. The Greek historian Herodotus describes in the following manner how the first systematic geometrical observations were made. The inundations of the Nile, bringing with them its fertile silt, would obliterate the boundaries between properties; each year these boundaries had to be delineated anew. This task, which would be troublesome even to a modern surveyor, had to be carried out rapidly and justly. It used to be performed by specialists, whom later the Greeks referred to as "harpenodapts", i. e. ropetyers—since, apparently, their main tool was the geodetic rope (today we use the geodetic tape). More detailed information about the proceedings of the harpenodapts has not been preserved. There is no doubt, however, that constant work on the same subject must have led to a considerable familiarity with geometrical figures and to the revelation of various laws. The harpenodapts were held in high esteem by their contemporaries. Democritus, the fifth-century Greek philosopher, boasted that nobody, not even the Egyptian harpenodapts, could excel him in the art of drawing lines, testifying thereby that in his time the Egyptians still ranked high as the most skilful geometers.

In the other countries of the East, in Babylonia and Assyria, geometry was also cultivated, though perhaps to a lesser extent. During the past twenty-five years, numerous mathematical texts in the cuneiform characters have been deciphered. It appears from them that the Babylonians had developed to a considerable extent the theory of equations; they were, for instance, able to solve quadratic equations. They also knew and were applying Pythagoras' theorem, the discovery of which should consequently be placed several centuries before the birth of Pythagoras. It is impossible to decide whether Pythagoras rediscovered it or whether he merely took it from Babylonian tradition and transplanted it in Greece. What most interests us here is the fact that geometry had already started in the Mediterranean countries and penetrated from them to Greece long before the Greeks became active in that field. The credit for introducing this science was attributed by Greek historians to Thales of Miletus (sixth century B.C.), but, when we bear in mind the lively trade-relations between Greece and Egypt, he certainly cannot have been its only propagator. In the sixth century B.C. began the development of Greek geometry, shortly to flourish magnificently.

What was the standard of the Greek geometry in the sixth century? We lack records from this period. We have to depend on the accounts of authors who were writing much later and on indirect deduction. The former, for example, attribute to Thales the discovery of the theorem relating to the isosceles triangle and the vertical angle theorem, which suggests that Greek knowledge of that period was confined to simple basic principles. On the other hand, certain works have survived which bear witness to the skilful application of constructional methods. There still exists today a tunnel dug in the sixth century B.C. through a hill on the island of Samos by an architect called Eupalinus. During the construction of this tunnel, which is two-thirds of a mile long, the adits were started on both sides of the hill and met in the middle with an error that scarcely amounted to a few yards. This is an impressive result when we remember that theodolites and other instruments now used were unknown in those days. We do not know Eupalinus' procedure; he must at any rate have been acquainted with numerous geometrical properties and have been able to measure angles accurately, and to calculate accurately the difference in level between the ends of his tunnel. At all events, he proved a master of the practical application of geometry. We gather from all this that the Egyptians and their successors, the Greeks of the sixth century, had collected a considerable knowledge of geometry, especially of those aspects of it which were of practical importance in building and similar occupations.

Into all this crude and empirically collected material the incomparable Greek genius introduced logical order, transforming a conglomeration of scattered facts into a compact science which was capable of deducing one theorem logically from another. This process, of course, lasted over many generations.

It seems that the first steps in this direction were taken by Pythagoras and his pupils, known as the Pythagoreans. A Greek historian (Eudemus, as quoted by Proclus) tells us: "Pythagoras has transformed geometry by formulating all-embracing principles and developing theorems by means of pure abstract argument". Tradition considers Pythagoras to have been the first to seek clarity in the concepts used and refers to him as the originator of the idea of definition. In the Pythagorean school (fifth and sixth centuries B.C.) abstract views were conceived for the first time; namely that a geometrical line has length but no breadth, that a circle is a line all of whose points are equidistant from a fixed point, and that a tangent to a circle is that straight line which has one point only in common with it. This standpoint, that a tangent to a circle has only one point in common with it, is already a far cry from any conclusions that could be reached by direct experimental observation of real straight lines and real circles. In the fifth century B.C. it was subject to vehement criticism from Protagoras, who pointed out that a real tangent to a real circle has in common with it by no means one point, but in fact a definite segment (Fig. 1). Protagoras accused these geometrical notions of fictitiousness; he indicated that the geometry deals with objects which do not and cannot exist, i. e. with arbitrary and preposterous inventions. A science, he said, ought to examine reality—that which exists in fact. Protagoras' objections were by no means shallow ones and it is worth while to examine this question in detail.

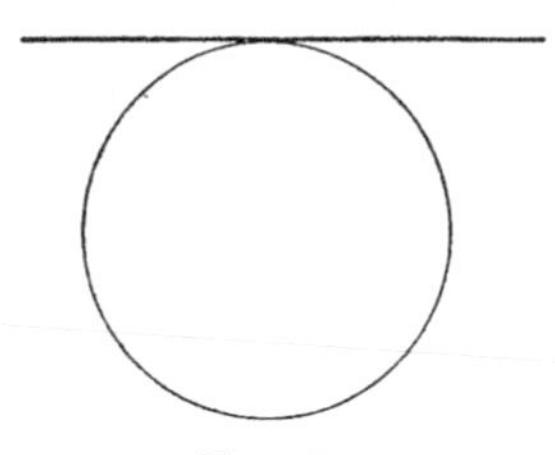

FIG. 1

A straight line drawn on a piece of paper is in fact a strip. A strip, admittedly, of minuscule width, but a strip nevertheless. The same applies to a drawn circle. Now these two lines are tangent when one strip overlaps the other as in Fig. 2.

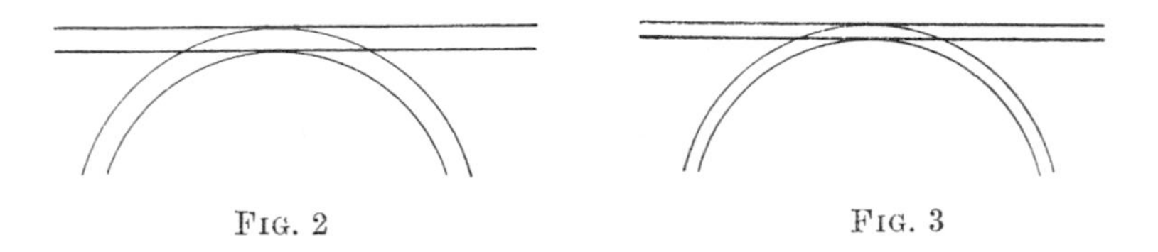

FIG. 2 FIG. 3

These strips have, then, a common part which is obviously not a point but which has a certain "length". One might think at first glance that this fact is due simply to the imperfection of the draughtmanship and ought to vanish, or at least to be considerably diminished, with the use of more subtle drawing instruments. However, the matter is not so simple. Let us look at Figs. 2 and 3. The strips in Fig. 3 are much thinner than those in Fig. 2.

Nevertheless the common length of the strips has remained almost the same. Now if we imagine that these strips are drawn thinner and thinner we must assume at the same time that our sense of sight becomes more acute if it is to perceive these minute objects—so that the common length of the two strips will appear greater. Similarly if we examine a strand of spider's-web through a magnifying-glass we will see it quite distinctly although it may be invisible to the naked eye, but at the same time its length will also apparently increase. In other words, if we were to observe a circle and its tangent made of the finest strands of cobweb, and if our eye were able to distinguish these strands, their common length would not appear to be so small. The size of this common part should not be estimated by comparison with a fixed unit of length, a centimetre for instance, but by comparison with the width of the "strips"—that is, we should consider the ratio of the common length of the lines to their width. A piece of elementary calculus work gives here an unexpected result. It appears, in fact, that the ratio of the common length of a circle and its tangent to the "width" of the lines by no means diminishes as we draw them thinner and thinner, but distinctly increases. We cannot therefore refute Protagoras' objections by blaming the drawing instruments; we cannot assert that the tangent theorem will work with greater exactitude as we use better materials, nor can we maintain that the properties of the figures of our "practical" geometry will tend more closely to those of our "abstract" geometry as we use more and more perfect methods of draughtsmanship.

As with the circle and its tangent, we meet with the same difficulties in other geometrical problems. We state, for example, that two straight lines intersect in one point, or in other words that two arbitrary intersecting straight lines define a point. Every draughtsman will without doubt contend that two perpendicular straight lines factually determine a point; but this breaks down for straight lines that form an angle of less than ten degrees, such straight lines do not "determine" a point (Fig. 4).

FIG. 4

No—such straight lines cannot be employed for the precise definition of a point. It can be shown that if two perpendicular straight lines are taken as cutting at a "point", the straight lines in Fig. 4 cut at seven "points", and matters are not altered when the lines are drawn thinner.

In conclusion: "theoretical" geometry cannot be considered as the limiting case of the "real" one as the sizes of points and the widths of lines decrease. It is a representation of reality which is simplified in another way. Protagoras was right: there is a difference between real facts and the postulates of theoretical geometry.

§ 2. Plato

As may be inferred from the title of one of Democritus' works there developed during the fifth century B.C. a discussion around the criticism of Protagoras. We know, however, nothing about its course. But we may guess from several passages in the works of Plato, and especially of Aristotle, who was continually returning to the subject of the circle and the tangent, that the matter had aroused real interest and endless argument. Later ages, under the spell of the triumphant development of theoretical geometry and the extraordinary usefulness of its applications, somewhat slided over the fundamental speculations of Protagoras, but in the fifth and fourth centuries they were certainly not treated lightly. We possess no records by means of which we might trace the evolution of Greek opinion, but it seems likely that the objections of Protagoras and the desire to refute them exercised an essential influence on the views of Plato (fourth century B.C.).

Generally speaking, it is difficult and in many ways controversial to characterize Plato's doctrine. Plato makes his points in a poetic and picturesque manner, using numerous suggestive comparisons; he wishes at times to draw the reader into his frame of mind, into his ardour for research, of which his dialogues are so full, rather than to communicate to him accurately and methodically the results of his enquiries. Moreover, there is nothing stiff and academic in Plato's doctrine; ideas conceived in one dialogue are modified in others—not only modified, but sometimes also caricatured, mocked and cast aside to make room for new ones. That flexibility of his views which reflects the constant evolution of thought steadily searching for the truth, but never satisfied with the results, is the reason why commentators hold to this day contradictory opinions as to Plato's theses is most principal matters. It is simply impossible to formulate these theses precisely without violating them, distorting and changing their colours.

Fortunately, for our purposes, there is no necessity to discuss the whole of Plato's philosophy, and it suffices to present his views on the relationship between theoretical and empirical geometry.

Of course, this question is only a small part of a more extensive problem, but concerning this small part Plato's standpoint, as set forth in the dialogue *Republic* and in the *Letter VII*, is quite clear and leaves no room for serious doubt. Plato admits that "the circle drawn or manufactured by man is far from our notion of a circle", since "such a circle coincides in every portion with a straight line"; he does not deduce from this, however, that geometrical theories deal with objects that have no real existence. For the subject-matter of these theories does not consist of drawn or manufactured circles and straight lines, but of ideal circles, ideal straight lines and ideal triangles, or in his own words of the "ideas" of a straight line, of a triangle or of a circle. These are by no means fictions, vain toys of the human mind, but have an objective existence independent of the human imagination, and are everlasting and unchanging. "Beyond the limits of the stars", says Plato, "exist pure ideas, without shape or colour, intangible and invisible not fixed in sensible particulars but free and independent". We would very much like to include this beautiful sentence in a poem extolling a mahtematical paradise, in which dwell ideal polygons, circles, spheres, regular icosahedrons and other geometrical figures; among them all striding portentously the logarithm, surrounded by a retinue of square, cube and fourth roots... In scientific considerations such fabulous pictures seem a little odd, but we must bear in mind that in the dawn of science all theories about the universe were based not exclusively on observation and argument but were conglomerates in which the conclusions of enquiry and cool logical speculation were overlaid by poetic fancy. Centuries were to pass by before man learnt to be cautious and critical in science.

With this in mind we may appreciate more fully Plato's granting of an independent existence to ideal geometrical concepts. He imbued geometry with the character of a true science dealing with existing reality; the realistic Greek mind was apt to recoil from a science concerned merely with intellectual inventions.

If we accept the real existence of the world of geometrical forms, we must ask two questions. The first is "Why in fact to bother to examine it?" and the second "How to do it?"

The need for study and research was obvious to Plato. Only ideal geometrical forms are governed by simple laws and only they can claim to have everlasting and invariable existence. The objects of this world reproduce these circles and straight lines only approximately and are influenced by accidents, and at the same time the relationships between them are only hazy reflections of the relationships between perfect forms. The reality we perceive thus stands as a representation of those relationships between perfect objects which play, as it were, the rôle of "pre-examples". Plato does not explain how it comes to pass, but makes his point clear by means of an allegory about prisoners in a cave, with a wall before them. At the entrance to the cave a fire is burning, and people are moving freely in front of flames. The passers-by and the objects they carry throw shadows on the wall. The prisoners may see the vague and vacillating shadows and gather from them how the objects are looking, but the shadows reproduce reality imperfectly and so the prisoners are incompletely informed.

Quite so are things in geometry. Geometrical relationships between the real objects under discussion are like shadows of the relationships between perfect beings. It is with these latter that man should try to be acquainted, for only they are primary and essential. Only they can give us true knowledge.

As for signposting the way which leads to this knowledge, we turn to the problem arising in the second of our questions. Here we discard our fabulous pictures and enter the field of the methodology of scientific work which, according to Plato, requires immense effort and indefatigable perseverance. In the scientific examination of an object Plato distinguished a number of succeeding steps. Let us take the example of a circle. The first step is the name-giving, the second the definition—that all points on a circle are equidistant from its centre. The third step consists in the producing of an image of the circle, accessible to the senses, e. g. in the drawing of it; such a picture is something "totally different from the idea of a circle". The fourth step is the scientific recognition, the embracing of the object by reason, the acquisition of an objectively true notion of it. This is achieved in a mental process, not appealing to the sense-images and without recourse to language. It is this activity of the mind which brings us closest to the essence (the idea) of the object. He who does not pass through the above four stages will never reach this essence.

Protagoras' sophistical criticism and Plato's dreams about the real world of ideas led to distinct progress, the significance of which must not be underestimated.

First of all it has now been acknowledged once and for all that geometrical concepts are not the same as their real counterparts in the material world; thus an empirical observation of a certain phenomenon is not and cannot be sufficient foundation for its acknowledgment as definite "pure geometrical" truth.

Almost without exception records of the older Greek mathematics have not survived to the present day, but we may guess that its arguments were a mélange of strictly logical deductions with references to facts which may have seemed obvious but which had not yet been precisely examined. Such a guess would be supported by the only

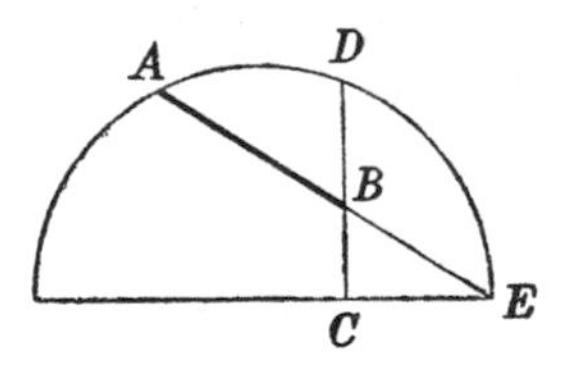

Fig. 5

longer fragment from the fifth century B.C. which has come down to us, namely, a tract on half-moons by Hippocrates. In it we find a series of precise arguments, and among them a reference without comment to the following problem: to construct a segment AB of given length with ends on a given semicircle and straight line CD respectively (Fig. 5), which when continued passes through the given point E on the diameter of the semicircle.

It seems plausible that the possibility of constructing such a chord was assumed intuitively by Hippocrates; the given segment AB, when A moves round the semicircle and B slides on CD, will assume a position where the continuation of AB passes through E.

It is possible that not only Hippocrates concerned himself with this problem, but it is unlikely that it was solved otherwise than by trial and error. For two hundred years later we find Apollonius of Perga also treating it, which would not have been the case if the solution had already been known for centuries.

In the Platonic school such a mixture of logical argument and appeal to intuition would not do; not so much, perhaps, because of the greater logical demands as on account of the cardinal principle: that the world of empirical reality is something lower, something baser, and cannot bear upon the higher, sublime world of perfect geometrical forms.

Platonic conceptions have produced an effect upon the development of deductive thought in yet another respect: to be sure, the final aim of scientific knowledge is, according to them, the comprehension of the idea of an object, but the preliminary condition for this is the formulation of its definition. Therefore the search for the correct definition became the daily preoccupation of the Platonic school, and Plato himself devoted much effort to this in his later dialogues. There has survived in historical literature (in the *Lives of the Philosophers* by Diogenes Laertius) the following characteristic story. Man, as he used to be defined by the Platonic school, was a mortal being, two-legged and featherless. This definition became quite popular in fourth-century Athens. The philosopher Diogenes, the *enfant terrible* of the Athenian community, plucked a live cock and took it along to the Platonic school, known as the Academy, saying "Here is your man". Diogenes' objection was not dismissed as mere facetiousness but accepted in all seriousness. In fact the plucked cock was, by definition, a man. As the result of this the definition was extended by addition of the words: "and having smooth nails".

In this curious and unexpected way the metaphysical phantasies of Plato contributed a great deal to the analysis of logical argument and led to an increasing interest in mathematics and logic. The Platonic Academy produced many eminent mathematicians. One of them, Theudios, wrote a text-book of geometry. In the Academy, also, logic was first formulated as a distinct branch of science, later systematised by the Academy's most distinguished pupil, Aristotle.

§ 3. Aristotle

Aristotle was of a cool, penetrating mind, extremely erudite—the first "scholar" in the modern sense of the word—critical and not given to flights of poetic fancy. So it is no wonder that he rejected the theory of the real existence of "ideas" and wrote bluntly: "To say that ideas are patterns of things and that things contain within themselves something of ideas is idle talk." He devoted many pages of his books to his fight against the Platonic concepts which, one gathers, were widely accepted.

Having rejected the real existence of ideal straight lines and circles Aristotle was forced to face the question: "What, then, are the objects with which mathematics deals?"

Quite certainly, they are not simply the objects of this world, for "none of them is of the sort that mathematics is interested in". Every kind of knowledge deals with objects that are perceptible to the senses, which are studied in physics. In this case, says Aristotle, "we must consider wherein the mathematician differs from the physicist.

For physical bodies contain planes, solids, lengths and points—which are what the mathematician investigates... The mathematician studies these figures, not qua limits of a natural body... He separates them since they can be separated in thought." "He investigates things after eliminating all sensible qualities such as weight, lightness, hardness and softness, also heat and cold... leaving only the quantitative and continuous... He investigates them in relation to nothing else."

Thus the objects of geometry, straight lines, circles, etc., have in Aristotle's philosophy, lost the real existence given them by Plato and have become the products of a complicated process of thought. Aristotle puts it concisely: "Mathematical objects we consider as the results of abstraction; physical objects have further properties".

Aristotle describes somewhat briefly how general concepts arise, by storing in the memory the features of objects which are similar to one another, i. e. how the process of this abstraction proceeds. He is concerned chiefly with its results. The human mind carries out this process of abstraction by collecting together the simple, general characteristic features of the real observed objects. The resulting concepts give the "essence" of things and geometry gives a picture of reality, one-sided but correct. Mathematical objects, to be sure, have no separate existence from the real ones, continues Aristotle, but "we treat them as separate". Thus, in spite of the utter difference in their views, the disciples of Aristotle were talking the same language as the pupils of Plato. Even today we have not swerved from their course; we say that Napier has "discovered logarithms" in the same way that we say an entomologist has "discovered" a new and beautiful species of butterfly.

Aristotle's studies of the structure of "proving" (or as we should say nowadays, "deductive") knowledge were of the greatest significance for geometry. Analysing the process of argument—basing it, it seems, to a great extent on a contemporary text-book of geometry—he became aware that not everything can be proved, since each argument must rest on previous information, which may in turn be based on yet earlier evidence. But since this process cannot be carried on *ad infinitum*, it is necessary to draw a line somewhere. Thus knowledge must be founded upon some principles which are taken for granted without proof. These are of two kinds.

One kind expresses certain general laws which find application in many sciences, e. g. that the differences of equals are equal (a favourite example of Aristotle), which works as well in geometry as in arithmetic. Aristotle lays stress on the importance of axioms, especially those relating to the concept of quantity (another example: that two quantities equal to a third one are equal to each other). We do not know if the credit for realising the need for the precise formulation of such obvious laws must go to Aristotle or to his predecessors. In any case, these axioms have now passed into all text-books of geometry and have been acknowledged as a foundation of mathematics.

Other principles of thought, are, to Aristotle, definitions in which the human mind describes concepts which are to reflect the reality. Such a definition need not be, as Plato would have it, a stage of a mental process whose culmination is the inward grasping of the idea of an object, but it must formulate the "essence" of the object under consideration. This essence must not be understood metaphysically, but actually and descriptively. They are, argues Aristotle in his *Analytics*, those attributes of an object which are peculiar to it but whose whole cannot be predicated of any other object; this whole of necessity constitutes the essence of the object. This idea is illustrated by the example that man is a mortal being, two-legged and featherless. These features are peculiar to all individually-existing men, but not to other creatures.

Definitions are the foundations of knowledge. If we

omit nothing which is peculiar to a certain object, says Aristotle, we can prove anything about it which is capable of being proved.

Nowadays we attach perhaps somewhat less importance to definitions than did Aristotle, but must admire the acuity of his further observations in which he points out that it does not suffice to formulate the definition of an object, but that one must also prove the existence of the object which is being defined. What is the point of saying what a *tragelaphus* (1) is, if the beast does not exist? Let us explain his idea by the following example:

It is not sufficient to define parallel lines as non-intersecting straight lines lying in the same plane, but one must also prove their existence. This we do, as it must

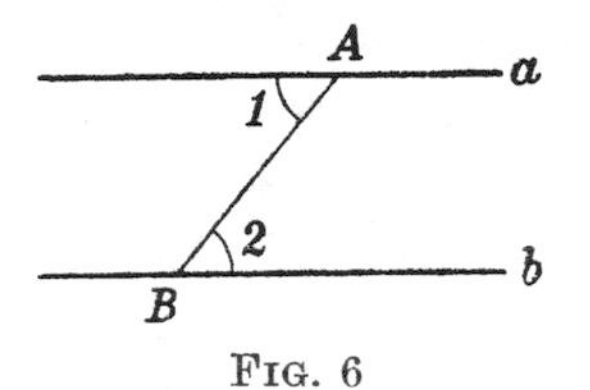

FIG. 6

have been done by the Greeks in the fourth century B.C., by constructing the lines a and b (Fig. 6) forming equal alternate angles *1* and *2* with a third line AB, and then proving that they cannot have any point in common.

There are also, according to Aristotle, certain concepts whose existence cannot be proved, since each stage of reasoning refers to a knowledge of objects whose existence has already been grasped, and continuing in this direction one comes to a stage where there is nothing on which to find support. The existence of such an object—Aristotle gives unit and magnitude as examples—should be taken for granted (postulated) and the post-Aristotelean geometry lays down explicitly the "postulates" which are necessary,

(1) A mythical creature, half deer and half panther.

e. g. that there "exist" circles with arbitrary centres and arbitrary radii.

Amateur mathematicians would consider such postulates to be pointless, and ask flippantly how a circle could not exist, when we can draw it with a pair of compasses. Our postulate, however, formulates abstractedly precisely that it is possible to move one leg of the compasses round the other.

The statement that every science rests on principles which are "unprovable" and originate from empirical observations has effectively opposed *a-priori-ism.* The requirement that every definition be accompanied by the proof of the existence of the object defined has been accepted by science once and for all. The concept of the postulate derived therefrom has played a fundamental rôle in the further development of mathematics.

Nevertheless, not everything Aristotle tells satisfies us. He divides knowledge into definitions and proofs (as we should say, definitions and theorems). Yet what is known by the first and what by the other? Aristotle attempted to draw a line between the spheres of action of proofs and definitions, but without convincing results. As he says, "definitions are principles of proofs", and the human mind forms these as a certain extract of reality. Further details as to how in fact to make this extract of reality are missing. In particular, the question whether definitions and postulates of existence contain all that geometer does extract from his observation of the world, and whether these, together with general axioms, suffice for the establishment of this science, is not answered.

It seems that Aristotle was of this opinion.

§ 4. Euclid and the axiom on parallels

Thirty or forty years after the work of Aristotle, i. e. about 300 B.C., were written *The Elements* of Euclid, an incomparable masterpiece of systematic, deductive

Greek thought, giving in thirteen books the geometrical and arithmetical knowledge of the times. We cannot here evaluate the immense influence exerted by *The Elements* on the development of science as a whole and not only in the field of mathematics, nor can we analyse their contents. It is sufficient to say here that the geometrical books of *The Elements* coincide almost exactly with the usual school course of geometry, and Bertrand Russel tells that when he was young, Euclid was the sole acknowledged text-book of geometry for boys in Britain. For the time being we are interested only in Book I, and especially in the paragraphs dealing with the foundations of geometry.

The exceptional precision of Euclid's thoroughly logical mind enabled him to realize the fundamental fact which escaped the notice of Aristotle. The action of defining and reaching the "essence" of concepts, together with general axioms—generally speaking those which refer to the properties of quantities—does not suffice for a logically correct development of geometry. One must, in addition, accept without proof certain laws, certain specifically geometrical axioms. Aristotle's postulates of existence are such axioms, but it turned out that in geometry it is necessary to refer not only to the definitions of concepts but to certain relationships between them. To us, who are accustomed to the study of relationships between objects, there is nothing surprising in this, but to Aristotle knowledge consisted first and foremost of the study of the characteristic attributes of these objects.

We shall not quote all the axioms listed by Euclid; we are now aware that his list is not complete, that is, that he did not formulate all the necessary axioms, although he noted the most essential ones. Here, however, are two:

1. *A straight line may be drawn through two points.*

2. *If a straight line c intersects two other straight lines a and b and makes with them two interior angles on the same*

side (1 and 2 in Fig. 7) *whose sum is less than two right angles, then a and b meet on that side of c on which the angles lie.*

In Fig. 7 the lines a and b intersect on the right hand side of the line c.

Why Euclid needed this axiom, known as the *axiom of Euclid*, we shall see below; its relation to experimental data will be discussed later on.

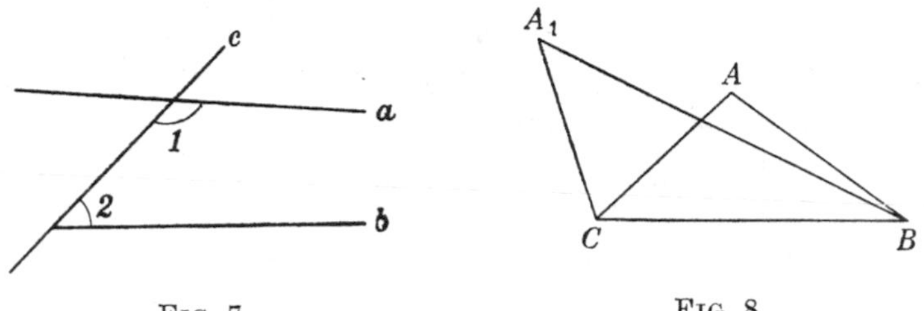

FIG. 7 FIG. 8

The first twenty-eight paragraphs of *The Elements* develop the theorems on congruent triangles, on the isosceles triangle, on the construction of perpendiculars. We also find here the theorem that the exterior angle of a triangle is greater than either of the interior and opposite angles, and some other properties of triangles, for instance that the sum of two sides of a triangle is greater than the third side. These paragraphs also state the theorem that if we do not alter the lengths of two sides of a triangle but increase the angle between them, the third side will become longer: i. e., in Fig. 8

$$A_1B > AB.$$

All these theorems are proved without referring to the axiom of Euclid. They are, as we say, independent of it.

Euclid was quite conscious of this independence, as is obvious from the arrangement of his exposition.

In the paragraph 27 Euclid demonstrates the method of constructing straight lines with no points in common, i. e. parallel lines.

To do this it is necessary, as pointed out above, to construct on AB (Fig. 9) equal angles *1* and *2*.

The straight lines *a* and *b* forming equal interior alternate angles *1* and *2* with AB are parallel. In fact, says Euclid, if they intersect, for instance, at the point K on the right of the figure, angle *2* would be an exterior angle of the triangle ABK but would be equal to the interior angle *1*, which is impossible according to the theorem mentioned above.

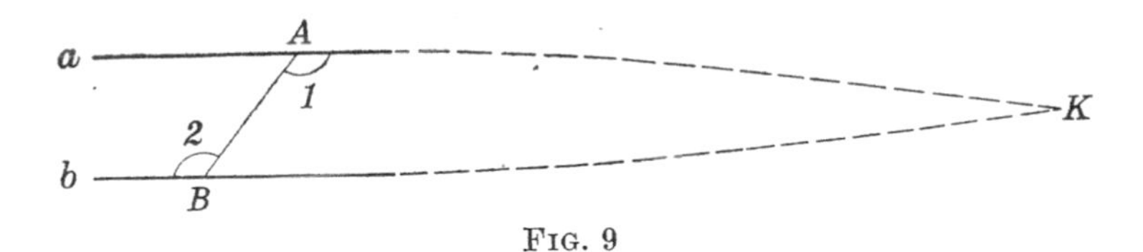

Fig. 9

As we see, this argument is not based on the axiom of Euclid. This axiom is not necessary until the beginning of paragraph 29, in which we find the proof of the following theorem: *If two straight lines are parallel, then their transversal forms with them equal interior alternate angles*, i. e. in Fig. 10,

$$\text{if } a \| b, \text{ then } \measuredangle 1 = \measuredangle 2.$$

Euclid argues as follows: $\measuredangle 1 + \measuredangle 3 = 180°$, since *1* and *3* are adjacent. If $\measuredangle 1$ were greater than $\measuredangle 2$, the sum

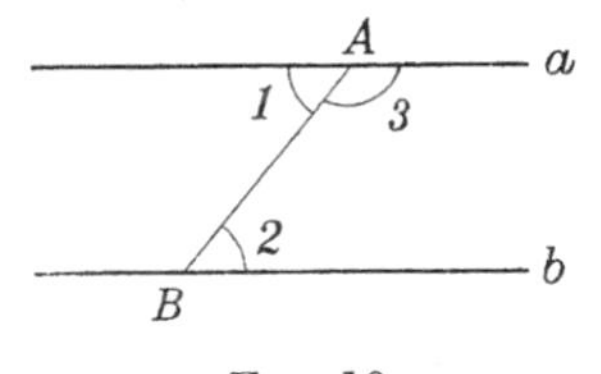

Fig. 10

of $\measuredangle 2$ and $\measuredangle 3$ would be less than 180°, whence by the axiom *a* and *b* would intersect. But since they are assumed to be parallel, we have the contradiction; *1* cannot be greater than *2*. Similarly it cannot be less than *2*, so the two angles must be equal.

So we see that the axiom of Euclid was necessary to prove that interior alternate angles formed by two parallel lines with their transversal must be equal.

This theorem is basic for proving many other theorems of geometry. Let us mention one or two of them.

1. *The sum of the angles of a triangle is equal to two right angles* (180°).

We construct a line e parallel to the base of the triangle through its apex A (Fig. 11).

Then $\sphericalangle 4 = \sphericalangle 1$, since they are equal alternate angles formed by e and BC with AB. Similarly, $\sphericalangle 2 = \sphericalangle 5$. But

$$\sphericalangle 4 + \sphericalangle 3 + \sphericalangle 5 = 180°,$$

$$\sphericalangle 1 + \sphericalangle 3 + \sphericalangle 2 = 180°.$$

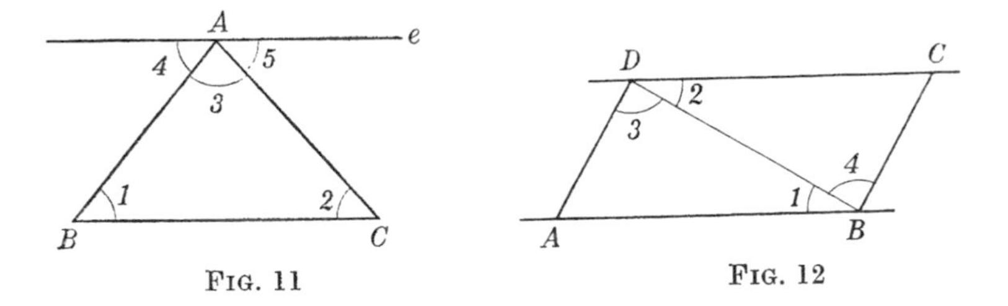

FIG. 11 FIG. 12

2. *Through a point A not lying on the straight line b there passes one and only one straight line parallel to b.*

3. *Parallel segments contained between two parallel straight ines are equal* (Fig. 12).

We get the equality $AD = BC$ by showing that the triangles ADB and BDC are congruent; this follows precisely because the alternate angles *1* and *2* are equal (the same applies to *3* and *4*).

4. A corollary from the above theorem is that *if parallel lines intersect two straight lines a and b and cut off from*

one of them equal segments, then the segments cut off from the other line will also be equal, i. e. in Fig. 13,

$$\textit{if segment } I = \textit{sgm } II, \textit{ then sgm } III = \textit{sgm } IV.$$

This theorem is the basis for the so-called theorem of Thales, which runs as follows: *segments formed in a straight line a by a number of parallel lines are proportional to those formed in b*, i. e.

$$\textit{sgm } I : \textit{sgm } II = \textit{sgm } III : \textit{sgm } IV.$$

The entire theory of similar triangles is based on this theorem, since its starting-point is the figure obtained by

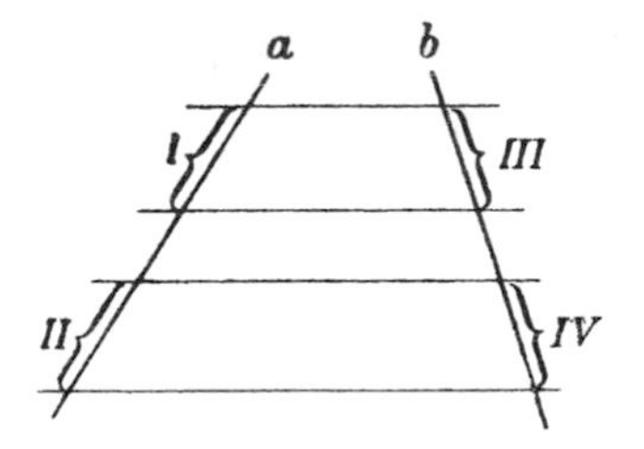

Fig. 13

intersecting the triangle with a parallel to one of its sides. Consequently, all the relationships in a triangle and all plane trigonometry follow from the theorem of Thales.

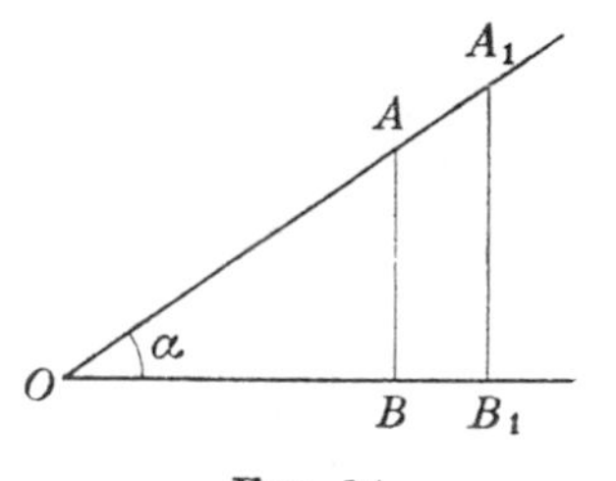

Fig. 14

Which theorem is the most essential in ordinary trigonometry? Obviously, that one which enables us to refer to the ratio of the perpendicular AB to the hypotenuse OA (Fig. 14) as *the sine of the angle* α, and to the ratio of the base OB to the hypotenuse OA as *the cosine of the angle* α.

The reader, perhaps, will protest and say: "Why should we not call the ratio "$\frac{AB}{OA}$" sine if we want to? After all, it depends only on us". Yet this is not so. When we introduce the term "the sine of the angle α" we are presupposing thereby that the ratio depends only upon the magnitude of α and does not depend on w h i c h point of the line OA has been chosen as A; in other words we have assumed that

$$\frac{AB}{OA} = \frac{A_1B_1}{OA_1}.$$

This equality follows from the theorem on similar triangles, and in turn from the theorem of Thales and finally from the axiom of Euclid.

5. The theory of the circle is also dependent upon the axiom of Euclid on that part of it which deals with the angles at the circumference and at the centre.

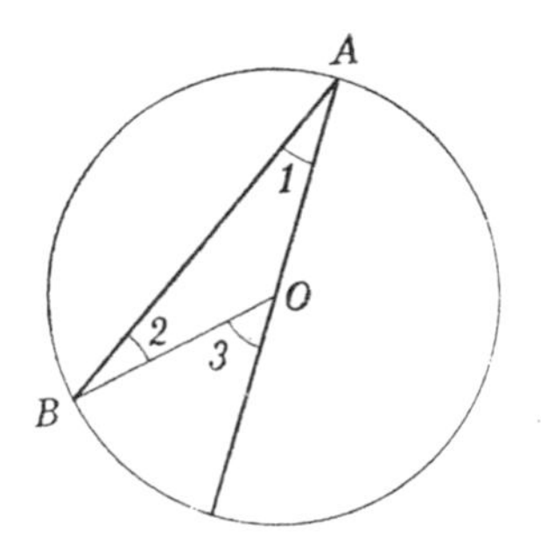

FIG. 15

In fact, to show that the angle at the circumference is half of the angle at the centre, when both have the same arc as their base, we consider the isosceles triangle AOB (Fig. 15).

The angles *1* and *2* are equal, and their sum is $180^\circ - \sphericalangle AOB$ by the theorem on the sum of the angles of a triangle, that is, directly from the axiom of Euclid. But $180^\circ - \sphericalangle AOB = \sphericalangle 3$, so $\sphericalangle 1 + \sphericalangle 2 = \sphericalangle 3$ and $\sphericalangle 1$ is half of $\sphericalangle 3$.

Therefore the theorem on the geometrical locus of a point from which a given segment is visible at a right angle also follows from the axiom of Euclid.

6. Let us finally mention the theorem that *parallel lines are equidistant*, i. e. in Fig. 16, that

If the straight lines a and b do not intersect, then the distances of all points on a from the other line are equal.

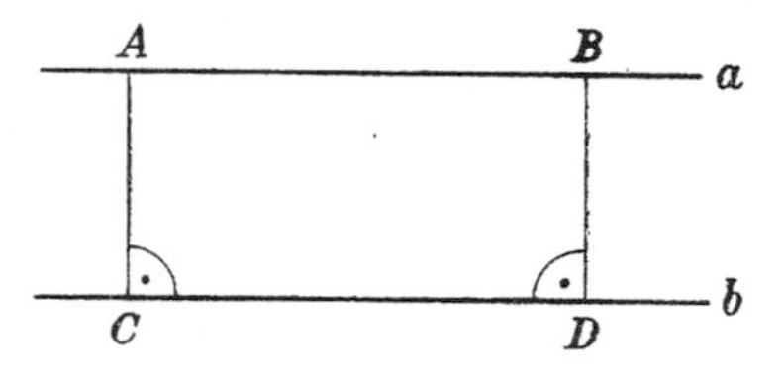

Fig. 16

Such distances are the lines AC and BD perpendicular to b. They form equal alternate angles with b, whence they are parallel.

We now have parallel segments between parallel straight lines and it follows by the theorem quoted at No. 3 above, thus indirectly by the axiom of Euclid, that $AC = BD$.

Conversely, points equidistant from b and on one side of it form a straight line parallel to b.

The above review demonstrates how great is the rôle played by the axiom of Euclid in geometry—how little would be left if we no longer accepted the truth of this axiom in our school-books. Naturally, Euclid was not the first to use it; the fact that a transversal intersecting two parallel lines forms equal interior alternate angles was, in all probability, known for a very long time before. It is unlikely that Eupalinus, without it, would have succeeded in digging his tunnel on Samos. In any case, the axiom of Euclid is by no means a deeply-hidden truth. Direct observation shows that if two actual straight lines a and b form the angles *1* and *2* with the line AB (*2* in Fig. 17 is a right angle) and the sum of $\measuredangle 1$ and $\measuredangle 2$ differs from 180°, then a and b will intersect. It is easy to discover by experiment that if this sum were 178° the distance of the point of intersection of the lines from B, in Fig. 17, would be about three yards. This experiment would provide a little more difficulty, but would still be practicable, if the sum of *1* and *2* were 179° 50′, since the distance would then be 34 yards.

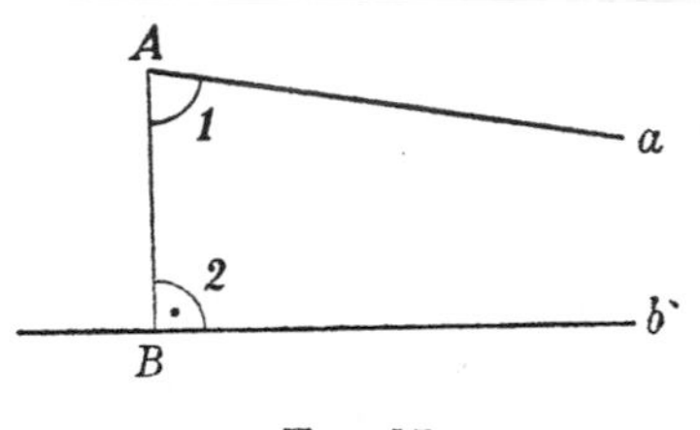

FIG. 17

The properties of parallel lines, of parallelograms etc., were known in the Pythagorean school and there is reason to believe that the proofs of many theorems, such as that on the sum of the angles of a triangle, were the same as they are now. Nevertheless, the theory was not entirely satisfactory. This we know from the writings of Aristotle, namely from his *Analytics*. There he criticises the procedure of those who would prove a certain property (A) on the basis of (B), which in turn is derived from (C), and finally conclude (C) from (A). Thus according to Aristotle, they assert that something holds because it holds. He reproaches the contemporary theory of parallels with this logical fallacy, this vicious circle. To Aristotle's pupils, for whom the *Analytics* were written, the above comments were clear, but we can only surmise how geometry used to be exposed in the fourth century B.C. Greek literature of later times indicated on several occasions the insufficiency of the following argument, which, as one may infer from the criticism, used to be applied:

Let the angle *2* in Fig. 17 be right and the angle *1* acute. Then, as is obvious and not difficult to prove, the line a will approach line b, if we move along a on the right hand side of A. Hence, a must finally intersect b.

It may be further deduced that the axiom of Euclid still holds in the case where *2* is not a right angle, but the sum of the angles *1* and *2* is less than 180°.

The above argument, stating that two lines approaching each other will necessarily intersect, must have been rejected the moment it was discovered that lines may approach nearer and nearer and yet not intersect. Such lines, for example, are the hyperbola with equation $y = 1/x$, which was already known to the Greeks as a section of a cone, and its asymptote, the horizontal axis (Fig. 18).

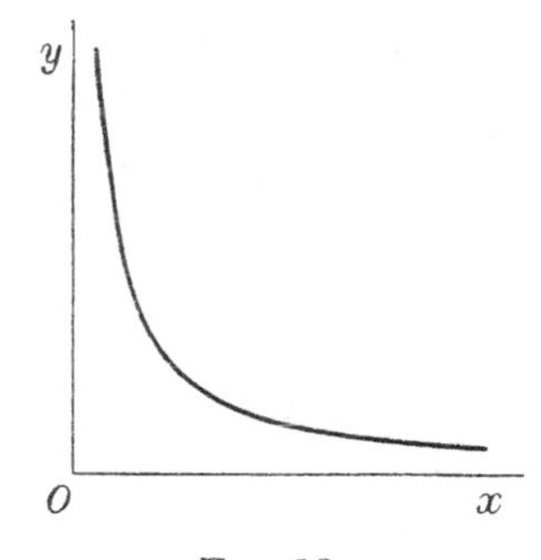

FIG. 18

The hyperbola comes infinitely close to the horizontal axis, but does not intersect it.

Thus the unlimited approach of two lines to each other by no means implies that they will intersect. In order to prove that the straight lines a and b will intersect (Fig. 17), it is not sufficient to show that they approach each other, but one must also appeal to other properties which are peculiar to straight lines but not to others, for example to hyperbolas.

Many readers will hasten to protest that the matter is clear by itself, "immediately evident", that straight lines cannot approach and not intersect. Thus saying, they arm themselves with their conviction of the truth of this theorem from intuition, that is from their experience of geometry; they assume the theorem as a self-evident truth, not deduced from other theorems. Which means that they accept the theorem as an axiom—just as Euclid did.

Aristotle's allusion probably did not apply to the fallacy mentioned above, since there is no vicious circle inherent in it. We may, however, find a vicious circle in the following argument, which is a simplified version of the reasoning of Geminus, a mathematician living some centuries after Euclid. It has been passed down to us by an Arab commentator.

It is easily seen that the axiom of Euclid (let us call it (A) in the terminology of Aristotle's criticism on page

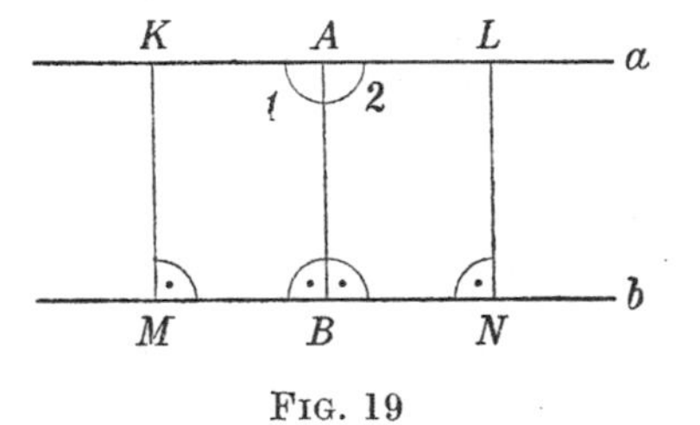

FIG. 19

32) follows from the theorem (B), that if a straight line is parallel to one of two parallels a and b it will also be parallel to the other. We shall deduce (B) from the theorem (C), that the straight line a is everywhere equidistant from the straight line b, or in other words that all segments with one end lying on a and perpendicular to b are equal.

Let us take $BM = BN$ and let KM, AB, LN be perpendicular to b (Fig. 19). The quadrangles $ABMK$ and $ABNL$ are congruent, since BN and BM are symmetrical about AB, whence KM and LN as perpendicular to them and equal (by (C)) are also symmetrical about AB. Hence, the corresponding angles *1* and *2* of both quadrangles are equal. These are adjacent, therefore right.

So far, so good, but we must now justify (C). True, we have done this on page 31 (the theorem at No. 6), but we were then basing our argument on the equality of interior alternate angles, and consequently on the axiom of Euclid (theorem (A)). The vicious circle has now been closed in conformity with the criticism of Aristotle: the truth of (A) has been made dependent on the proof of the truth of (A). For the sake of historical accuracy we should mention that Geminus did not base his argument on the theorem (C), stating that the points lying on a straight line a parallel to a straight line b are equidistant from b, but on its converse, that points equidistant from b and lying on the same side of it form a straight line not cutting b. The essential factor is, of course, that all such points form a straight line, and not some curve. Geminus, and quite a number of his followers, tacitly assumed this, that is, they considered it self-evident, a necessary property of straight lines, and by no means a new axiom.

We note in passing that points equidistant from a certain line l do not always form a line of the same shape as l. In Fig. 20 is drawn a parabola and a number of points equidistant from it; these points quite obviously do not lie on another parabola.

The fact that points equidistant from a straight line form another straight line is a characteristic property and we might repeat the remarks made above when discussing lines which approach but do not meet.

This investigation has given a positive result: namely, that the axiom of Euclid and the theorem according to which points equidistant from a straight line and on the same side of it lie on another straight line are equivalent statements in the sense that one of them can be deduced if the other is accepted as an axiom. Geminus did not realize this, and imagined that his exposition of geometry dispensed with the axiom of Euclid.

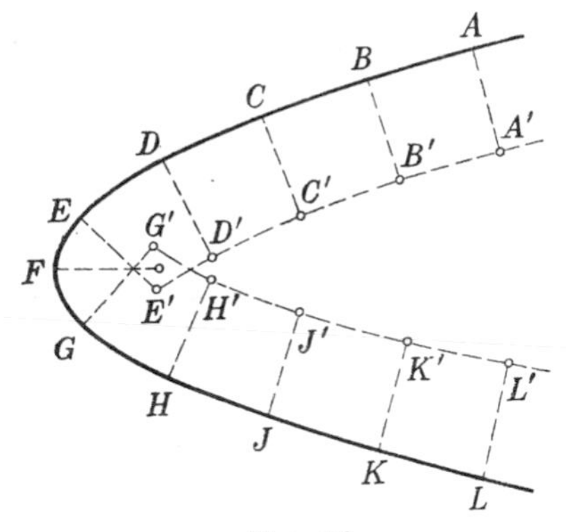

FIG. 20

Examining this fallacy of Geminus—our Arab commentator mentioned above repeats the former's arguments with applause—we must admire all the more the critical acuity of Euclid, who was not led astray by what appeared to be a proof and who considered it necessary to assume an axiom—rightly bearing his name.

Euclid was fully aware of the significance of his standpoint: the theorems of Book I of *The Elements* were arranged, as we have mentioned, in order to delay as long as possible the introduction and application of the axiom, even though its earlier use would have simplified some proofs. His aim was methodical—to lay down all that could be proved without appealing to the axiom on parallels, or in other words, to segregate that part of geometry which is independent of this axiom. This part of geometry is nowadays sometimes referred to as *absolute geometry*—a rather odd term, originating from Bolyai. Euclid used no such term, but realized to some extent the importance of dividing absolute theorems from the rest. One should not gather from this that he considered the latter to be less reliable than the former; he was, rather, governed by a logician's instinct and pigeonholed together those theories which rested on common or analogical foundations. These tendencies appear frequently in modern mathematics, especially in algebra, and do give a certain tone to the science.

As we have said, it was logical requirements which decided Euclid to assume the axiom on parallels, he thought this axiom be necessary for the correct development of geometry. This point of view had several consequences; apart from anything else it was an opposition to Aristotle's methodological directions, for Aristotle wished to found geometry on general axioms such as "two quantities equal to a third are equal to each other" or "the part is smaller than the whole" and on definitions which fixed the meaning of geometrical concepts. The results of observing the outward world of geometrical forms are expressed, according to Aristotle, in the choice of suitable abstract notions, copies, so to say, of reality in thought. The postulates of existence form the bridge to connect these notions with reality. Geometrical theorems are deduced systematically and consecutively from the essence of these concepts formulated in definitions. Putting it concisely and somewhat simplifying, Aristotle wished to build up the science of space on the general laws of human thought and on the definitions. His idea is not missing in modern science; many mathematical theories are developed in this way, this procedure is used, for instance, in four-dimensional geometry.

Having introduced his axiom Euclid did, to a certain extent, break away from the narrow Aristotelean pattern. He found, and future ages fully agreed with him, that there was nothing in the essence of the concept of a straight line which would force one to assume that two straight lines forming with a third unilateral interior angles with a sum less than 180° must necessarily intersect. Hence, this property should be formulated separately. Therefore it is from Euclid that we have the view that geometry is to be founded on concepts and axioms relating to them (that is, not general axioms, but specifically geometrical axioms)—a view wider than that held by Aristotle. The process of arriving at the knowledge of the real world lies not only in a skilled formulation of concepts but also in the fixing of the principal relationships between them. The theorems of geometry are then deduced from the axioms without a repeated appeal to intuition.

The above views do need any special comment, since nowadays they are generally acknowledged and taught in schools. Nevertheless it would be worth while to mention that over the course of centuries they met with lack of comprehension and with opposition. In 1733 still the Italian mathematician G. Saccheri tried to prove, in a dissertation entitled *Euclides ab omni naevo vindicatus* (1), that two straight lines cannot intersect in two points and also that there is a straight line which passes through two given points. These were not, of course, proofs in the present meaning of the word but somewhat primitive appeals to intuition in which Saccheri considered an arbitrary curve between points A and B (Fig. 21), rotating it from the left-hand side of the points A and B to the right-hand side, and finally bringing both curves closer to each other until they coincided. Saccheri did not grasp the depth of Euclid's conceptions but was a shrewd man widely gifted. He used to play simultaneously three games of chess by memory, that is without looking at the board—and successfully too. In the dissertation quoted he has a proof of the axiom of Euclid (two proofs, in fact, but both wrong) which was, to him, the chief flaw in the Elements, and in doing so he noted and correctly proved an interesting theorem from absolute geometry.

(1) "Euclid cleared of all stain".

We mentioned Saccheri's attempts not because they have played any great rôle in the history of mathematics but in order once more to pay homage to the sharpness

FIG. 21

of Euclid's intellect. Not only fortune, but also opinion is variant. What Saccheri considered a fault in Euclid's exposition we now take to be one of his chief merits.

§ 5. Attempts to prove the axiom of Euclid

Saccheri's standpoint, as presented in the preceding section, was extreme. The necessity for taking not only general axioms as a base for geometry but also some specifically geometrical ones was by and large acknowledged, and axioms like "there exists a straight line which passes through two given points" caused no objections. They were not, after all, a far cry from the Aristotelean postulates of existence. Matters were different with the axiom on parallels. Almost from the very moment when *The Elements* appeared until the nineteenth century (over two thousand years!) this axiom continually aroused opposition and many attempts were made to rid geometry of it. There is something deeply moving in the epic of these heroic strivings towards ideal scientific perfection—disinterested effort directed solely by the love of knowledge.

In these endeavours it is possible to distinguish two main trends, which crossed, however, quite frequently.

First, it was attempted to replace the axiom of Euclid by another one equivalent to it. Such an axiom is, as we have seen, the one "that all points equidistant from a straight line and lying on the same side of it also lie on a straight line".

In the seventeenth century the Englishman Wallis showed that the axiom of Euclid is equivalent to the existence of triangles of different sizes but with equal corresponding angles—i. e. to the possibility of magnifying the triangle without changing its angles. In other words, if the axiom of Euclid were not true similar triangles could not exist, or, more generally, similar figures altogether, and we would be unable to scale plane objects, i. e. make maps. Wallis considered that an axiom which states the existence of similar figures lays down a more characteristic geometrical property than does the axiom of Euclid, whence it should be given precedence. Finally, it was discovered in the eighteenth century (Saccheri and others) that *the axiom of Euclid is equivalent to the theorem on the sum of the angles of a triangle.* As we have

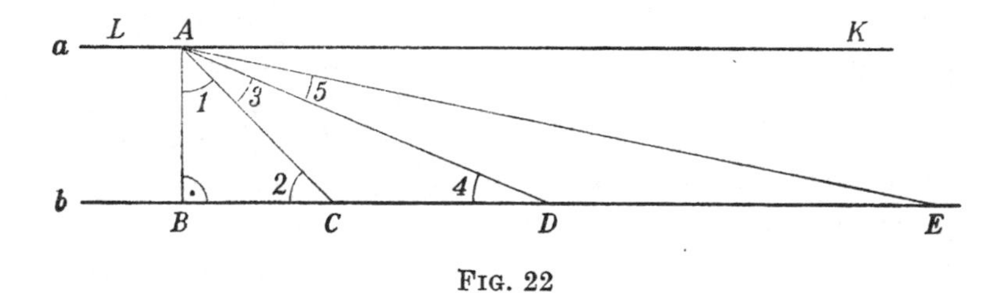

FIG. 22

seen, this theorem follows from the axiom of Euclid (the theorem at No. 1, page 28). Conversely, if we suppose that the sum of the angles in an arbitrary triangle is 180°, we can easily demonstrate the truth of the axiom of Euclid.

According to the comment made on page 34 it is sufficient to show that if the straight lines a and b are parallel (Fig. 22) and AB is perpendicular to b, then AB will be perpendicular to a.

Let us take $BC = AB$. $\triangle ABC$ is a right-angled triangle, the sum of its acute angles is $90^\circ = \mathrm{D}$ (we infer it from the theorem that the sum of the angles of a triangle is 180°). Since these acute angles are equal, we have

$$\sphericalangle 1 = \tfrac{1}{2}\mathrm{D}.$$

Let us now take $CD = AC$. The angle $ACD = 2\mathrm{D} - \sphericalangle 2 = 2\mathrm{D} - \frac{1}{2}\mathrm{D}$, whence the sum of $\sphericalangle 3$ and $\sphericalangle 4$ equals $\frac{1}{2}\mathrm{D}$, and these angles are equal, giving

$$\sphericalangle 3 = \tfrac{1}{4}\mathrm{D}.$$

Let $DE = AD$. As before we obtain

$$\sphericalangle 5 = \tfrac{1}{8}\mathrm{D}.$$

Continuing the process we have at the vertex A the following angles:

$$\tfrac{1}{2}\mathrm{D}, \tfrac{1}{4}\mathrm{D}, \tfrac{1}{8}\mathrm{D}, \tfrac{1}{16}\mathrm{D}, \ldots$$

The angle KAB is greater than $\sphericalangle 1$, than $\sphericalangle 1 + \sphericalangle 3$, than $\sphericalangle 1 + \sphericalangle 3 + \sphericalangle 5$, etc., i. e. it is greater than each term of the sequence

$$\tfrac{1}{2}\mathrm{D}, \tfrac{3}{4}\mathrm{D}, \tfrac{7}{8}\mathrm{D}, \tfrac{15}{16}\mathrm{D}, \ldots$$

which tends to D. So the angle KAB is equal to D or is greater than D:

$$\sphericalangle KAB \geqslant \mathrm{D}.$$

The same argument gives

$$\sphericalangle LAB \geqslant \mathrm{D}.$$

The sum of the angles KAB and LAB is 2D, whence owing to the above inequalities they are both right angles.

Thus the equivalence of the theorem on the sum, the angles of a triangle to the theorem of Euclid has been proved.

Secondly, it was attempted to deduce the axiom of Euclid from the remaining axioms. Several arguments which claimed to do this made use of facts which were obvious to their authors but were, when thoroughly examined, equivalent to the axiom of Euclid—that is, facts which could not have been proved themselves without the aid of that axiom. These arguments, then, contained the same logical fallacy as that of Geminus. Some of these "proofs" were quite ingenious and interesting. As an example we shall give here the proofs of the French mathematician Legendre, who inserted them into a school text-book which was in general use in the first decades of the nineteenth century. The poor schoolchildren were hammering away at false arguments until someone noticed the mistakes and when Legendre removed them from his book. But he still did not give up and found some more false proofs which were published as a special paper in 1833!

In the proof which we shall quote here Legendre tried to show that if we suppose the sum of the angles of a triangle to be greater than 180°, or to be less than 180°, we are faced with a contradiction.

1. Let us suppose that the sum of the angles of the triangle ABC (Fig. 23) equals $180° + \alpha$.

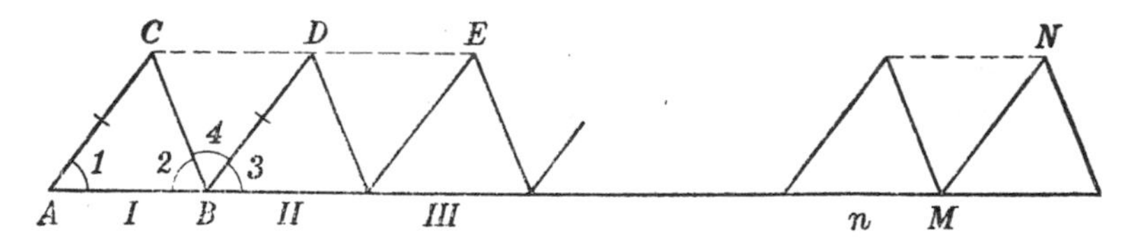

Fig. 23

We now construct on the same straight line AB a series of triangles congruent to ABC, as in the figure. Let us now join their apices. We then have

$$\measuredangle 1 = \measuredangle 3$$

but

$$\measuredangle 1 + \measuredangle 2 + \measuredangle C = 180° + \alpha \quad \text{(by assumption)},$$

whence

$$\sphericalangle 3+\sphericalangle 2+\sphericalangle C = 180°+\alpha.$$

On the other hand

$$\sphericalangle 3+\sphericalangle 2+\sphericalangle 4 = 180°.$$

Comparing this formula with the preceding one we get

$$\sphericalangle 4 < \sphericalangle C.$$

The ticked sides of the triangles ABC and CBD are equal, and side BC is common. The angle C in the first triangle is greater than the corresponding angle 4 in the second. Whence, from the absolute theorem mentioned on page 26 (Fig. 8), we obtain

$$AB > CD,$$

that is to say,

$$AB-CD > 0.$$

Let us now take a sufficiently large integer n so that $n(AB-CD)$ will be greater than $2.AC$, i. e.

$$n.AB-n.CD > AC+AC.$$

Let us consider (Fig. 23) n consecutive triangles like ABC and n more like BCD. MN is the side of the last one.

Since $n.AB = AM$ and $n.CD$ is the length of the polygonal line between C and N, namely the line $CDE\ldots N$, we may write the last inequality as follows:

$$AM-CDE\ldots N > AC+MN \quad \text{(since } AC = MN\text{)}$$

or

$$AM > AC+CDE\ldots N+MN.$$

On the left-hand side we have the segment AM, on the right the length of the polygonal line between A and M. Thus the straight distance from A to M would be longer than the roundabout distance, which is impossible.

We come to a contradiction, so the sum of the angles of a triangle can never be greater than 180°.

2. Now let us suppose that the sum of the angles of a given triangle ABC is $180° - a$.

Let the angle A be acute.

Let us construct on BC a triangle BCD congruent to ABC so that the sides similarly ticked in Fig. 24 will be equal. Let us draw a straight line passing through the point D and intersecting the sides of the angle BAC at the points K and L.

Four triangles have been formed, marked in the figure by the numerals *I*, *II*, *III*, *IV*.

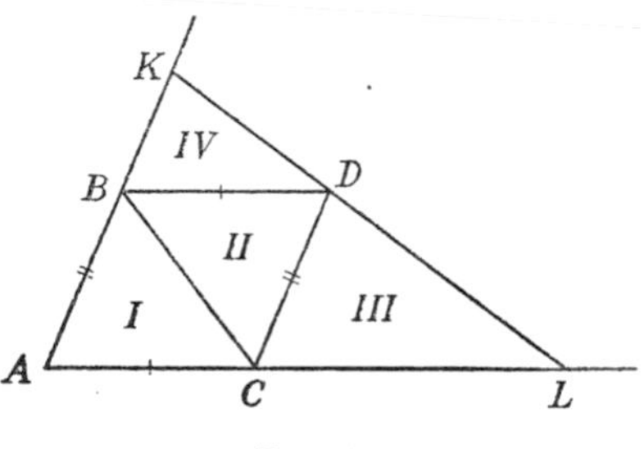

FIG. 24

The sum of the angles of each of the triangles *I* and *II* is $180° - a$, and the sum of the angles of *III* and *IV* cannot exceed 180°, according to the result of the previous argument.

Let us add all the angles of *I*, *II*, *III*, and *IV*. We obtain a sum not exceeding

$$(180° - a) + (180° - a) + 180° + 180° = 720° - 2a.$$

This sum consists of angles A, K, L and the three angles at B, C, D which are all equal to two right angles:

$$\sphericalangle A + \sphericalangle K + \sphericalangle L + 3.180°.$$

we get

$$\sphericalangle A + \sphericalangle K + \sphericalangle L + 3.180° \leqslant 720° - 2a,$$

and consequently the sum of the angles of the triangle AKL does not exceed $180° - 2\alpha$.

Applying the same process to triangle AKL we get (Fig. 25) a triangle AK_1L_1 whose angles have a sum not greater than $180° - 4\alpha$. Similarly we arrive at a triangle AK_2L_2, the sum of whose angles is not greater than $180° - 8\alpha$, etc. This process leads to a triangle the sum of whose angles is negative, since 2α, 4α, 8α, 16α, ... must eventually exceed 180°.

We have come to contradiction, says Legendre, so the sum of the angles of a triangle can never be less than 180°. Since this sum can be neither greater nor less than 180° it must be 180°. From this, as we know, follows the truth of the axiom of Euclid.

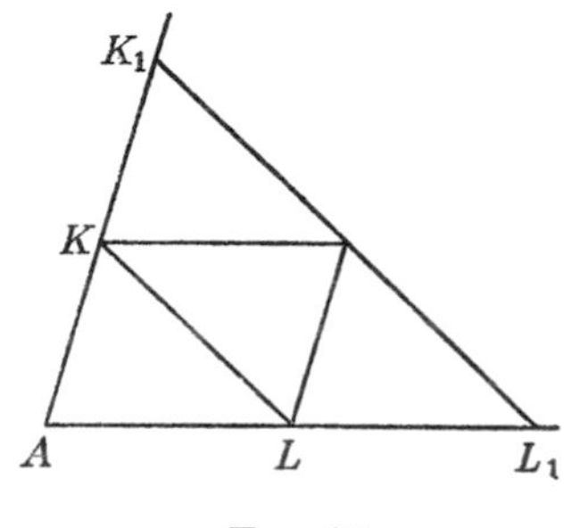

FIG. 25

The first half of the proof is correct. The theorem that the sum of the angles of a triangle cannot exceed 180° has been proved without appealing to the axiom of Euclid and therefore belongs to absolute geometry. The second half, however, is false. The reader is recommended not to read further but to try and detect the slip on his own. If he succeeds it will testify to his critical abilities and powers of observation and promise much for his further mathematical studies.

We will give away the secret. The weak point of the argument is the phrase: "Let us draw a straight line passing through the point D and intersecting the sides of the angle BAC at the points K and $L \ldots$". It is not difficult to find a straight line passing through a point inside the angle and intersecting one of its sides—it suffices to join D with any point on this side. But how can we be so sure that a straight line exists which passes through D and intersects both sides? This should be checked. A detailed analysis based on the axiom of Euclid shows that such a line is, for instance, the perpendicular dropped from D onto the bisectrix of the angle A. Therefore it follows from the axiom of Euclid that there exists a straight line with the indicated property. The second half of Legendre's proof shows that if we can find a straight line like this through any interior point of an angle the axiom of Euclid will be true. The mist has lifted a little. It now seems, much, no doubt, to the surprise of the reader, that the axiom of Euclid is equivalent to the theorem: *Through every point within an angle there passes a straight line which intersects both sides of the angle.*

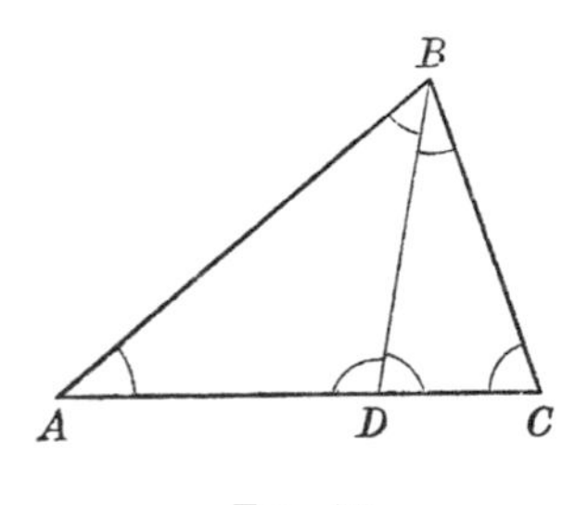

FIG. 26

The axiom of Euclid really seems like a magician with many guises, who takes us aback and deceives us.

We shall now deduce from theorem 1 a corollary enabling us to formulate still another statement equivalent to the axiom of Euclid.

Suppose that the sum of the angles in one triangle ABC is 180° (Fig. 26). Let us divide this triangle into the two triangles ABD and DBC the sums of whose angles are denoted by α and β respectively. Hence $\alpha+\beta$ is the sum of all the angles marked with arcs in the figure and is equal to the sum of the angles of triangle ABC, i. e. 180°, plus the sum of the two angles at D. Consequently

$$\alpha+\beta = 360°.$$

As we know, neither α nor β can exceed 180°, hence $\alpha = \beta = 180°$.

Further subdivision of either triangle will again produce triangles the sum of whose angles will be 180°, and therefore: *If the sum of the angles of a certain triangle is* 180°,

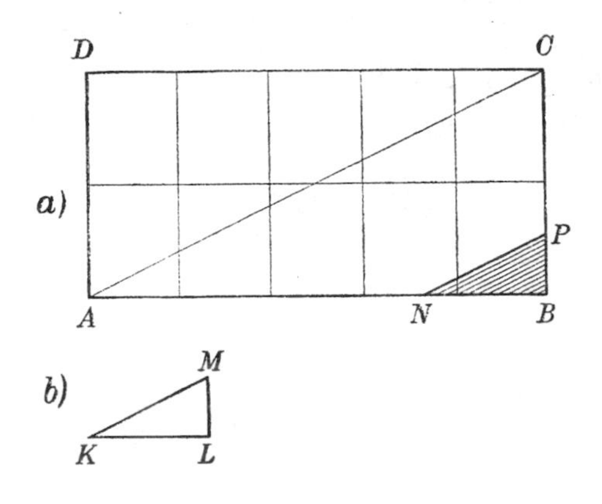

FIG. 27

then each triangle cut off from it will also have the sum of its angles equal to 180°.

We now affirm that the axiom of Euclid is equivalent to the following statement:

There exists at least one rectangle, that is a quadrangle with four right angles.

PROOF. If there exists one such rectangle, we may (Fig. 27) arrange identical ones next to each other so as to get a rectangle $ABCD$ with arbitrarily large sides. Let us consider half of $ABCD$—the triangle ABC. The sum of the angles of ABC is one half of the sum of the angles of $ABCD$, i. e. 180°.

In this case the sum of the angles of any right-angled triangle KLM is 180°, since KLM may be placed on ABC (provided that the latter has sufficiently long sides). Then, by the lemma given above, the sum of the angles of the triangle BPN is 180°, since BPN is cut off from ABC.

Now, since the sum of the angles of an arbitrary right-angled triangle is 180°, it follows from Fig. 28 that the sum of the angles of any triangle is also 180°, whence the axiom of Euclid holds.

It is clear that the converse also holds, and so the announced equivalence has been proved.

A slight modification of our argument would give the following, striking result: *If there exists at least one triangle the sum of whose angles is* 180°, *then the axiom of Euclid will be true.*

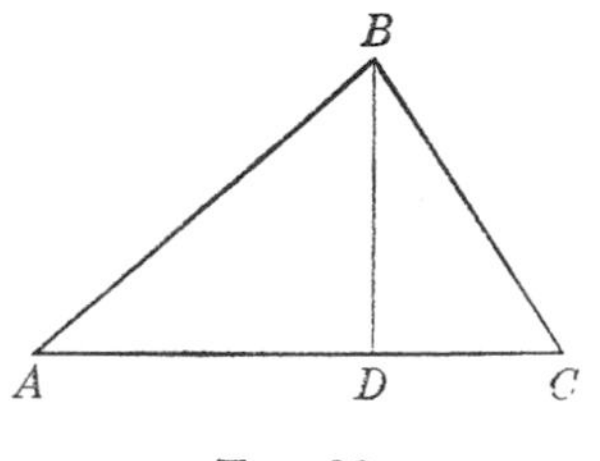

FIG. 28

In fact, if the sum of the angles of triangle ABC (Fig. 28) is 180°, then the same holds for the right-angled triangle ABD cut off from it. With two such triangles we can construct a rectangle and obtain thereby the conditions of the preceding theorem. Hence, the situation is extremely characteristic: *If the sum of the angles of at least one triangle is* 180°, *then the axiom of Euclid will hold and the sum of the angles of any other triangle will also be* 180°.

If the sum of the angles of at least one triangle is less

than 180°, then the sum of the angles of any other triangle will also be less than 180°.

These theorems have been proved without appealing to the axiom of Euclid and so belong to the realm of absolute geometry.

§ 6. The axiom of Euclid and the empirical knowledge

All our considerations up till now have referred to the logical structure of geometry. Their object was to discover whether the axiom of Euclid was indispensible to the structure of geometry or whether it could be deduced from other axioms; they are, so to say, logical amusements. No single mathematician entertained any doubts as to the truth of the axiom of Euclid. This "revolutionary" idea was first conceived by the great German mathematician Gauss in the first two decades of the last century. To him the question of whether the axiom of Euclid were true was of actual, physical significance; namely, it was a matter of whether real points and straight lines, as for example, those employed in land-surveying, would obey this axiom.

The last theorem of the preceding sections yielded a method of solving the question: one should measure the sum of the angles in any one triangle. If this sum appeared to be 180°, the axiom of Euclid would hold.

Gauss traced a triangle in the neighbourhood of Göttingen whose sides were thirty or so miles long and whose vertices were at the summits of mountains.

Then with the utmost precision he measured its angles. It appeared that the deviation of their sum from 180° lay within the limits of inevitable errors of measurement, and so it remained unsettled whether this sum was exactly 180° or differed from 180° by an amount less than those errors.

We shall now show that the failure of the attempt

could have been forecast in advance. To this end we shall consider some factors related to those dealt with in the last section.

Let us call the difference between the sum of the angles of the triangle and 180° the *defect* of the triangle.

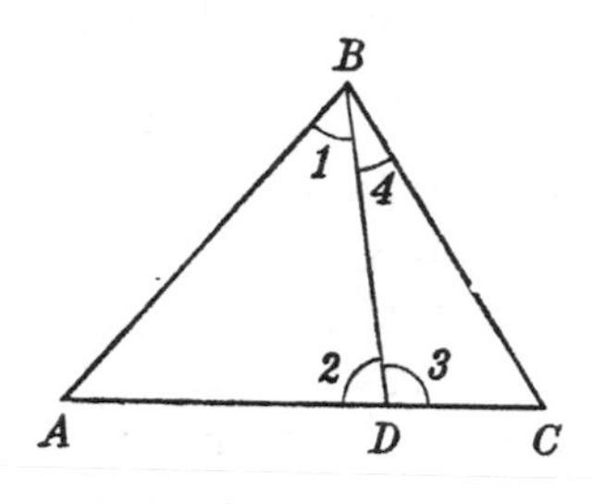

FIG. 29

Let us now divide the triangle ABC (Fig. 29) into triangles ABD and BCD. We have

$$\text{Defect } \triangle ABD = 180° - \sphericalangle A - \sphericalangle 1 - \sphericalangle 2.$$

$$\text{Defect } \triangle BCD = 180° - \sphericalangle C - \sphericalangle 3 - \sphericalangle 4.$$

Let us add these formulae.

$$\text{Defect } \triangle ABD + \text{Defect } \triangle BCD$$

$$= 360° - \sphericalangle A - \sphericalangle C - (\sphericalangle 1 + \sphericalangle 4) - (\sphericalangle 2 + \sphericalangle 3).$$

But

$$\sphericalangle 2 + \sphericalangle 3 = 180° \quad \text{and} \quad \sphericalangle 1 + \sphericalangle 4 = \sphericalangle B.$$

We obtain

$$\text{Defect } \triangle ABD + \text{Defect } \triangle BCD$$

$$= 180° - \sphericalangle A - \sphericalangle B - \sphericalangle C = \text{Defect } \triangle ABC.$$

The defect of the triangle ABC is equal to the sum of the defects of the triangles of which triangle ABC consists.

This theorem may be generalised for every partition of a triangle into triangular components; furthermore, for every partition of a polygon into polygonal components, provided, of course, that the defect of a polygon has been defined.

Let us now imagine that the triangle whose vertices are at the Sun, Earth and Mars has a defect of 1°, i. e. its angles add up to 179°. Let us divide it into smaller ones more or less of the size of that drawn by Gauss. The distances from each other of the Sun, the Earth and Mars amount to hundreds of millions of miles and it is easy to compute that the number of the component triangles will exceed a trillion. Thus the defect of a component triangle with a thirty-mile side would be something like a trillionth of a degree. Obviously, no instrument could detect such a tiny angle. This calculation, which is based on the theorem of defects, inclines one to believe that it would not be possible to meet with noticeable defects when measuring triangles on earth. As far as the earth is concerned the defect of a triangle is virtually zero and the sum of the angles of a triangle is 180°: in other words, the axiom of Euclid with all its consequences holds. Less microscopic defects would occur only in the triangles which occur in astronomical research. The biggest triangles accurately known are those serving for the determination of the parallaxes of fixed stars.

Let G be a fixed star, A and B two opposite positions of the Earth in its orbit round the Sun (Fig. 30). The segment AB is the diameter of the Earth's orbit. Its length is about 186 million miles. The angles GAB and GBA can be measured since their vertices lie on the Earth. Of course, we may choose A and B so that the angles are equal, which will happen if AB is perpendicular to GM. The quantity $\frac{1}{2}(180° - \sphericalangle GAB - \sphericalangle GBA)$, i. e. $90° - \sphericalangle GAB$ is called the *parallax* of the fixed star G. If the sum of the angles of the right-angled triangle GAM is 180°, then the parallax of the star will be $\sphericalangle AGM$, i. e. the angle which the radius AM of the Earth's orbit would subtend at G. If the sum of the angles of the triangle AGM were less than $180°$, the parallax of the star would not be equal but greater than the angle AGM.

It is evident that the defect of AGM, which is $180° - 90° - \sphericalangle GAM - \sphericalangle AGM$, is less than the parallax, which is $90° - \sphericalangle GAB$. Hence, we can estimate

$$\text{Defect} \triangle AGM < \text{parallax of the star } G.$$

The parallax of the star Sirius has been measured as $0{\cdot}38''$, and that of Vega as $0{\cdot}08''$. The corresponding defects are smaller than the parallaxes and are therefore

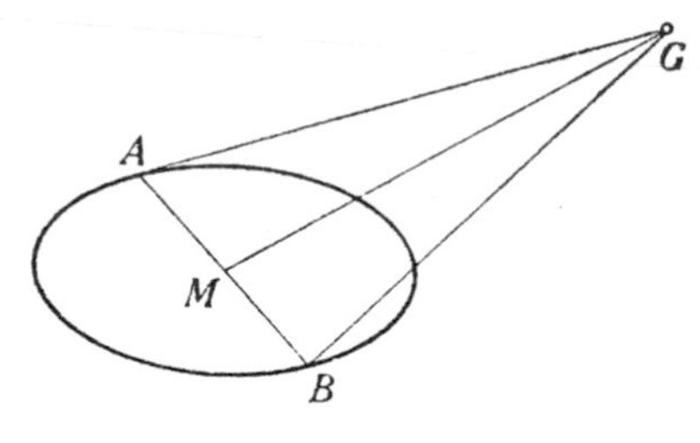

Fig. 30

very tiny angles. Only much greater triangles than those used when measuring parallax, which are very "narrow" indeed, could have larger defects; in the case of Sirius the side AG is about half a million times as long as AM.

No triangles bigger than parallax triangles can possibly be investigated and so the question of whether the defects of enormous triangles are zero, in which case the axiom of Euclid would hold throughout space, or not, has not been settled by the direct measurement of angles. Contemporary physical theories, viz. the theory of relativity, state that when we are dealing with very great distances the axiom of Euclid fails. We cannot here discuss these difficult matters nor consider the extent to which empirical data bear out the above view. In any case we must bear in mind the possibility that the sum of the angles of a triangle, although equal to 180° with overwhelming precision when dealing with terrestrial or even solar dimensions, may be less than this figure in triangles of "cosmic" sizes. Thus mathematics finds itself faced with the task of discovering the properties of triangles on the assumption that the sum of their angles is not 180°. What would, for instance, the theorem of Pythagoras look like then? How can the area of a triangle be computed? What is the relationship between the side of a right-angled triangle and the hypotenuse and one of its acute angles?

That system of geometry which is built up on all the axioms of ordinary geometry except that of Euclid and on the negation of the latter is known as *non-Euclidean geometry* (sometimes, for reasons which we shall not give here, *hyperbolic non-Euclidean geometry*). This science contains, clearly, all the absolute theorems, which remain the same as in ordinary, Euclidean geometry, but many theorems in addition which are different from the Euclidean ones: an example, which we know, is the theorem that there is no quadrangle with four right angles.

§ 7. The creators of non-Euclidean geometry

The first scholar who realised that non-Euclidean geometry might exist and who admitted it the right to exist was Gauss. He discovered many of the theorems of the new science, but printed none of his findings. Those which we know are gleaned from his note-book, which was published in later times. It is not known whether it contains all his research and his findings. With this reservation we may conclude, on the basis of the existing materials, that Gauss, absorbed in other work, did not come to very final results in the field of non-Euclidean geometry. This is true at any rate of the methods he used, if not of the actual contents of the theorems he discovered; indeed, in the most important of the preserved fragments he uses methods of differential geometry, a science based on differential calculus, whereas synthetic methods like those used in elementary geometry would much better fit the case. This would seem, to the present author, to be one of the reasons for Gauss's not communicating his findings to the world of science.

Meanwhile two young mathematicians, the Russian, Nikolai Lobatchevsky, and the Hungarian, János Bolyai, by a bold stroke of genius, developed the principles of non-Euclidean geometry, and settled nearly all of its essential problems.

Lobatchevsky (1793-1856), a professor at Kazan University, published his first paper *On the principles of geometry* in the 1829-30 numbers of a journal which appeared in Kazan but did not reach other countries.

Bolyai (1802-60), an officer of the Austro-Hungarian army, presented his discoveries, carried out independently of those Lobatchevsky's, in a paper entitled *Appendix scientiam spatii absolute veram exhibens* (1) that appeared in 1832, few years later than the publication of Lobatchevsky's work. Thus the priority of discovery must go to the latter, and non-Euclidean (hyperbolic) geometry is accordingly called also *Lobatchevskian geometry*. Incidentally, it is truly amazing to what extent the trains of thought of the two scholars were related; they were in essence both based on the properties of the *horosphere* (*vide* § 17, p. 116).

Both dissertations were ignored by the scientific world and brought their authors none of the acknowledgement which their independence of ideas, ingenuity of argument and perfection of results deserved. They were both aware

(1) "Appendix giving an absolutely true science about space". The term "appendix" derives from the fact that the dissertation appeared as a supplement to a text-book of mathematics written by Bolyai's father.

of the value and importance of their work. Bolyai wrote with pride to his father: "I have created a new world out of nothing". Both expected, and were entitled to expect a rightful appreciation. Both met with complete indifference or even, in the case of Lobatchevsky, with jeers from people who were somewhat narrow-minded and failed to comprehend what it was about. Lobatchevsky and Bolyai were both bitterly disappointed but reacted, however, in different ways. Bolyai, exasperated, closed his mouth and withdrew from scientific activity. Lobatchevsky took up the struggle for the triumph of his ideas; in publication after publication he doggedly justified his non-Euclidean geometry from every point of view and indicated its applications in the integral calculus in the hope that he would finally win comprehension and acknowledgement. He dictated his last work, *Pangeometria*, seriously ill, almost blind, but not giving up the struggle even in the last days of his life. The extraordinary steadfastness of spirit shown by Lobatchevsky during his twenty-five-year struggle in utter isolation has very few equals in the history of science. We must, however, admit that a university professor is able to preserve his independence more easily than a minor functionary like Bolyai.

In the sixties and seventies of the last century the concepts of non-Euclidean geometry spread and its creators, ignored during their lifetimes, were included in the Pantheon of the greatest scholars. This change of minds was certainly influenced by the publication of Gauss's correspondence, which was not sparing in its praise of Lobatchevsky and Bolyai. A more essential reason, however, was the natural evolution of scientific interest which took to questions of geometry, regarding them from new vantage-points. The celebrated works of Staudt appeared, analysing the principles of projective geometry, and Riemann's lecture *On hypotheses basic to geometry* made an immense impression.

With the air thus cleared, many scholars [1] turned to the problems of non-Euclidean geometry, filling in the still-existing gaps and refining its methods.

Of greatest significance was the lecture of Riemann. It created a very general science, known as *Riemannian geometry*, whose special, in fact very special, cases are the Euclidean and non-Euclidean geometries. This science exceeds both the scope of the present book and any possibilities of an elementary presentation, so we shall not discuss it further. Our aim is to give the principles of non-Euclidean geometry as they were formulated by Lobatchevsky and Bolyai, but with some simplifications, the most important of which are due to the recent Danish mathematician Hjelmslev [2]. We shall also give in detail—in honour of the centenary of Lobatchevsky's death in 1856—what is perhaps the most beautiful idea in his dissertation, the use of the properties of the horosphere to prove the fundamental theorem of non-Euclidean geometry. Nowadays other proofs of this theorem exist, but to our mind it is the original method of Lobatchevsky which is the most interesting.

---

(1) Beltrami, Klein, Lie *et al.*

(2) Theorem and transformation j of § 9.

CHAPTER II

THE PRINCIPLES OF NON-EUCLIDEAN GEOMETRY

§ 8. Fundamental assumptions

Lobatchevskian geometry, as we said in the previous chapter, differs from ordinary geometry only in so far as it rejects the axiom of Euclid, so that all the theorems of Euclidean, ordinary geometry which do not rest on this axiom, i. e. the absolute theorems, are equally valid in Lobatchevskian geometry. These are the theorems which are discussed in school text-books before the chapter about parallel lines—namely, those about congruency, symmetry rotations together with their various consequences, as mentioned in part on p. 26. Also, that part of the knowledge of the properties of the circle which deals with chords and their corresponding arcs, with tangents and intersections of circles is also "absolute" (for instance, the greater the chord the nearer it is to the centre of the circle). Then there are numerous stereometric theorems which do not depend on the axiom of Euclid. They include all those which refer to the perpendicularity of straight lines and planes, for example:

1. *All straight lines perpendicular at one given point to a given line lie in one plane.*

2. *A plane which passes through a straight line perpendicular to another plane α will also be perpendicular to α.*

3. The so-called theorem on three perpendiculars: *If a line a is perpendicular to another line b lying in plane α then the projection of a onto α will also be perpendicular to b.*

All the theorems of elementary school geometry whose proofs appeal neither directly nor indirectly to the properties of parallels or the axiom of Euclid are absolute theorems. On the other hand it would not be true to say that every theorem proved in school text-books by using parallel-line properties is not absolute, for it may happen that such a theorem might be proved in an alternative way, without making use of the properties of parallel lines. An interesting example of this is the theorem: *That all plane angles of the same dihedral angle are equal.*

We have (Fig. 31) two plane angles ABC and $A_1B_1C_1$ of the same dihedron, that is, angles whose sides are perpendicular to the edge BB_1 of the dihedron and lie on its faces; we wish to prove that these angles are equal. To do this it is usually pointed out that the lines BC and B_1C_1, perpendicular to BB_1, are parallel, and similarly AB and A_1B_1. From the theorem on parallels it is easily seen that our angles are equal.

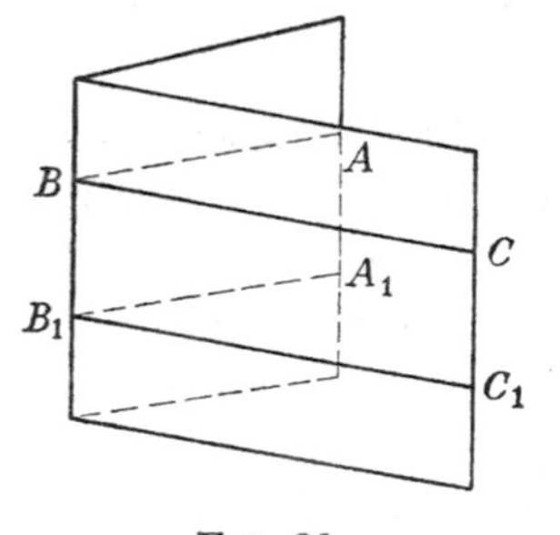

FIG. 31

Nevertheless, this theorem may be proved without appealing to the properties of parallels but using the properties of symmetry. The interested reader will find the proof in the Appendices to this book. Here we wish only to state that the theorem on the equality of plane angles of a dihedral angle belongs to absolute geometry.

Later on we shall appeal to several of the absolute theorems of the school text-book. In view of the above remarks they should give rise to no doubts—the latter would in any case vanish if any elementary handbook were consulted.

Let us now consider the consequences which follow from the negation of the axiom of Euclid, that is, from the assumption that not every pair of straight lines forming with a third straight line unilateral interior angles $\measuredangle 1$ and $\measuredangle 2$ whose sum is less than $180°$ must necessarily intersect.

In other words: There exist straight lines a and b which do not intersect, such that the sum of the angles 1 and 2 is less than $180°$ (Fig. 32).

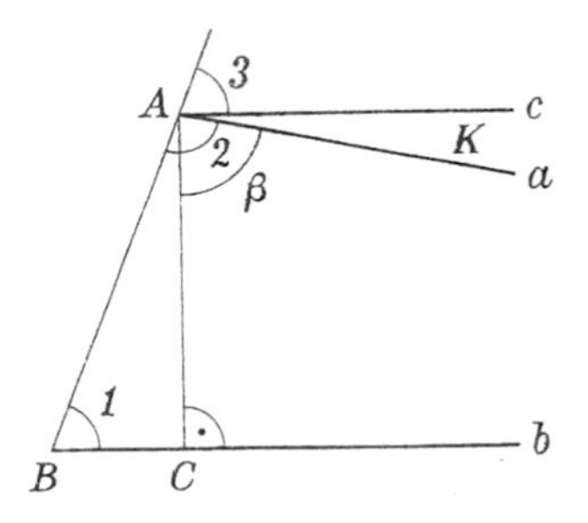

FIG. 32

If we construct the angle 3 equal to 1 on AB produced we get the straight line c which also does not cut b (see p. 26). Let us drop from A the perpendicular AC to b. Either a or c is not perpendicular to AC. Let it be a.

We reach the corollary: In the plane in which the axiom of Euclid fails, known as the *Lobatchevskian plane*, there exist two non-intersecting straight lines, of which one forms a right angle with AC and the other forms the acute angle β.

This may also be expressed as follows: *There exists an acute angle* β ($\measuredangle KAC$ in Fig. 32) *such that a certain perpendicular* (b in Fig. 32) *to one of its sides does not cut the other side.*

Let us draw the angle β separately and construct a perpendicular e from E to AC (Fig. 33a).

If we now shift the perpendicular from the position e to b, etc., the first perpendiculars $(e, f, \ldots)$ will cut a while further ones (e.g. b) will not.

The former perpendiculars are separated from the latter by a perpendicular HJ not cutting a (for, if it did (Fig. 33b), then a slightly lower perpendicular H_1J_1 would also cut a, and HJ is chosen so as to separate the upper perpendiculars which intersect a from the lower ones which do not).

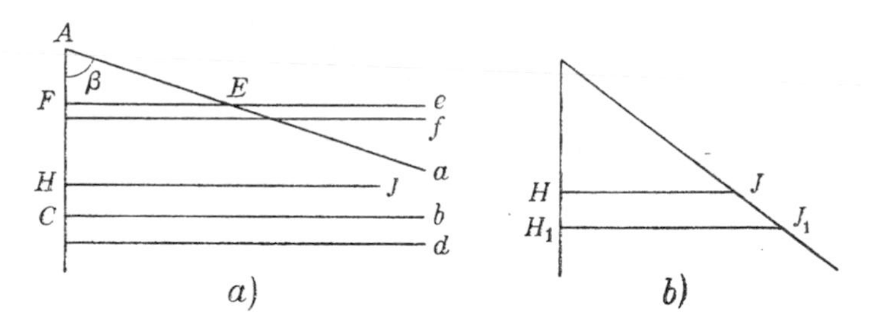

FIG. 33

Putting it briefly, constructing, from points of side a, perpendiculars to the other side of the angle β, we will never go beyond a certain line HJ, though we may approach it as closely as we please.

The foot of the perpendicular from the point E to the straight line is the projection of this point onto that line. Hence we see that *the projections of the points of one side of the angle* β *onto the other side fill the segment* AH (*not including the point* H) *but never go outside this segment.*

To our minds, which are accustomed to Euclid, this sounds paradoxical, yet it is an inevitable consequence of the negation of the axiom of Euclid. One may, incidentally, draw a figure in Euclidean geometry with a similar property. This is the angle formed by the half-line starting at A with the arc of the hyperbola which has asymptote HJ (Fig. 34).

The projections of the points lying on the arc of the hyperbola fill the segment AH but do not go outside it.

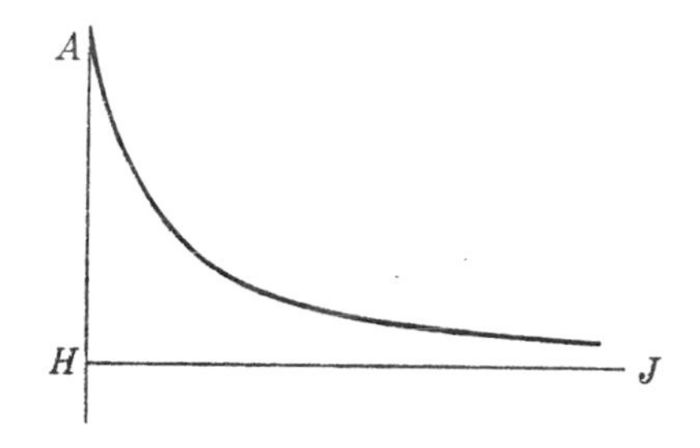

FIG. 34

Therefore, when speaking of the properties of angles in Lobatchevskian geometry, we shall illustrate them sometimes by means of Fig. 34.

§ 9. The theorems of Hjelmslev

First, let us recall the meaning of the term "transformation".

Let the straight line b be fixed. When we speak of transforming the points $A, B, C, \ldots$ symmetrically with respect to the axis b we mean finding points which are symmetrical to $A, B, C, \ldots$ with respect to the axis b, that is, so to speak, replacing A by A_1, B by B_1, etc. (Fig. 35).

Let us take the fixed point O (Fig. 36). If we replace A by A_1 lying on the half-line OA at a distance from O which is m times greater than OA, i. e. by taking $OA_1 = m \cdot OA$, we transform A by dilatation with respect to the centre O.

Similarly other points B, C are transformed into B_1, C_1.

Whenever we give a rule which assigns to each point A of a plane unique point A_1 of this plane (i. e. to different points A, B are assigned different A_1, B_1) we thereby define a certain transformation in this plane.

We have given above two examples of transformations: symmetry with respect to the axis b and dilatation with respect to the centre O.

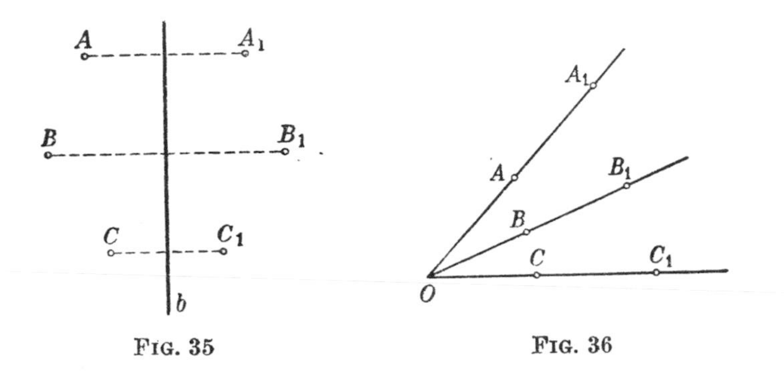

Fig. 35 Fig. 36

It is known that the symmetry and the dilatation transform a segment into a segment; however, only certain transformations possess this property. For instance,

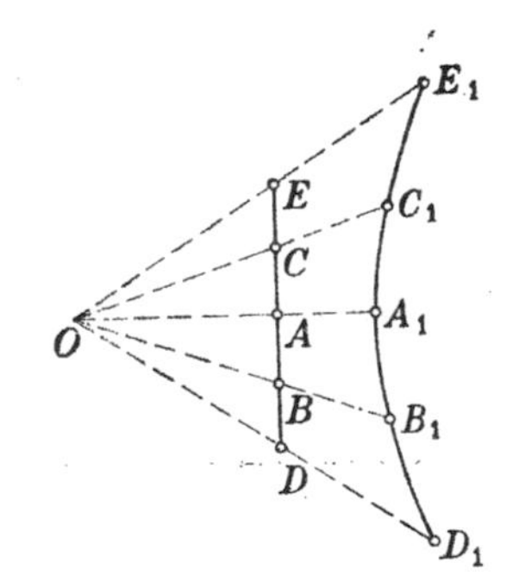

Fig. 37

if, in Fig. 36, we had taken OA_1 to be equal to $m \cdot (OA)^2$ instead of to $m \cdot OA$, straight lines would be transformed into curves, as may be seen in the figure—in Fig. 37, the unit of length is a centimetre and $m = 1$; some points A and the corresponding points A_1 are shown.

Not only the points in a plane may be subject to transformations. When making a map of a certain country, a part of the surface of our globe, we "transform" the points of the country into the points of the map. In this case we often say that we "map" instead of "transform".

Let us now discuss a certain absolute theorem which is going to play a fundamental rôle in the following pages. It is, interesting in itself, but surprisingly enough it is not generally given the attention which it deserves in the teaching of geometry. Let us consider two straight lines and such sequences of points on them $A_1, B_1, C_1, \ldots$ and $A_2, B_2, C_2, \ldots$ (Fig. 38) that corresponding segments of both straight lines will be equal, i. e.

$$A_1B_1 = A_2B_2, \quad B_1C_1 = B_2C_2, \quad A_1C_1 = A_2C_2 \text{ etc.}$$

We shall call them, in short, congruent sequences of points.

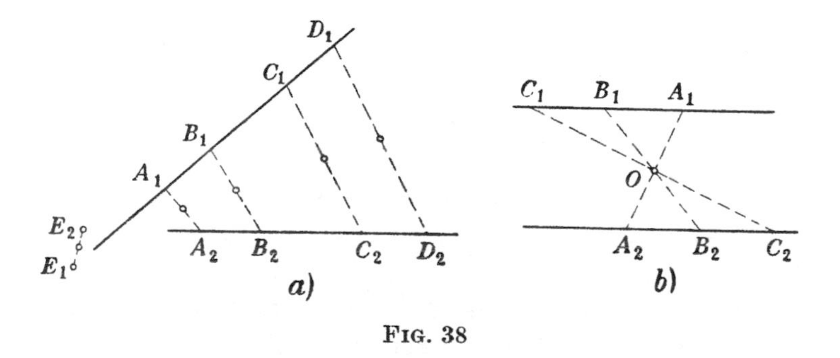

FIG. 38

Let us now join the corresponding points of both sequences by the segments A_1A_2, B_1B_2, etc. and find the centres of these segments. Looking at the picture we may guess the following theorem:

The centres of the segments joining the corresponding points of two congruent sequences of points lie on one straight line (are collinear).

This theorem has been known for a long time, but it was only Hjelmslev who discovered that it belongs to absolute geometry and has numerous applications. It is therefore called the *theorem of Hjelmslev*.

If the sequences of points are situated as, for example, in Fig. 38b, the centres of the segments A_1A_2, B_1B_2, C_1C_2... coincide in one point O but the theorem remains true. As may easily be seen, this special case only occurs when the sequences have a centre of symmetry.

We shall now prove the theorem of Hjelmslev.

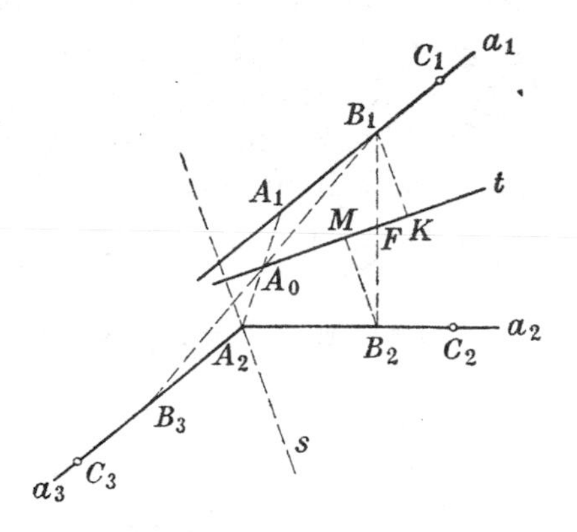

FIG. 39

Let A_0 be the centre of the segment A_1A_2 (Fig. 39). Let us rotate the whole plane round A_0 through 180°. A_1 will then be mapped into A_2, B_1 into B_3 and the line a_1 into a_3. The sequence of points thus obtained A_2, B_3, C_3... is symmetrical to the sequence A_2, B_2, C_2, ... about the bisectrix s of the angle $B_3A_2B_2$.

Thus if we first rotate the line a_1 round A_0 through 180° and then transform it by symmetry with respect to the axis s, we shall thereby map the series of points A_1, B_1, C_1, ... into A_2, B_2, C_2, ... Let us draw from A_0 a perpendicular t to s and from B_1 a perpendicular B_1K to t and then see what happens to t and B_1K with the above-defined rotation and symmetry.

The line t, when rotated through 180° round A_0, will be transformed into itself, since it passes through A_0.

Again, when transformed by symmetry with respect to s, it will reappear as itself, since it is perpendicular to s, the axis of symmetry.

Thus we see that both by rotation and by symmetry t is transformed into itself, which means that every point on this line will be transformed into another certain point on the same line; that is K will be transformed into, say, M.

On the other hand, as we know, by rotation together with symmetry B_1 is transformed into B_2 and so finally the segment B_1K into B_2M and the angle between B_1K and t into the angle between B_2M and t. Consequently both these angles are right and the segments B_1K and B_2M are equal.

Noting, further, that B_1 and B_2 lie on different sides of t we may easily see that the triangles B_1FK and B_2FM are congruent and, next, that F is the centre of B_1B_2, i. e. that the centre of B_1B_2 lies on t. Also, by the same argument, the centre of C_1C_2 lies on t, Q. E. D.

We shall now introduce a certain transformation in the plane which we shall call, for brevity's sake, *transformation h*.

Let us take a fixed point O and a fixed acute angle α (which, for the sake of argument, we measure clockwise). Let us connect O with an arbitrary point A and construct an angle α on the half-line OA with OB as its other side (Fig. 40).

The transformation h replaces point A by its perpendicular projection onto OB, i. e. by the point $A_0: A \rightarrow A_0$.

Similarly we map C into C_0, D into D_0, etc. Also, the arc DC of a circle with centre O into the arc D_0C_0 of another circle with the same centre.

The transformation h changes the sizes of figures but, as we shall show by means of Hjelmslev's theorem, it maps collinear points into collinear points—that is, what was "straight" before transformation will remain "straight" after it.

It is clear that points which lie on a straight line passing through O, e. g. those lying on OA, will be mapped into points lying on OA_0.

Now let us consider the points A, E, F, lying on the straight line a which does not pass through O (Fig. 41).

Let us rotate a round O through an angle 2α. We get the straight line a_1. The sequences A, E, F and A_1, E_1, F_1

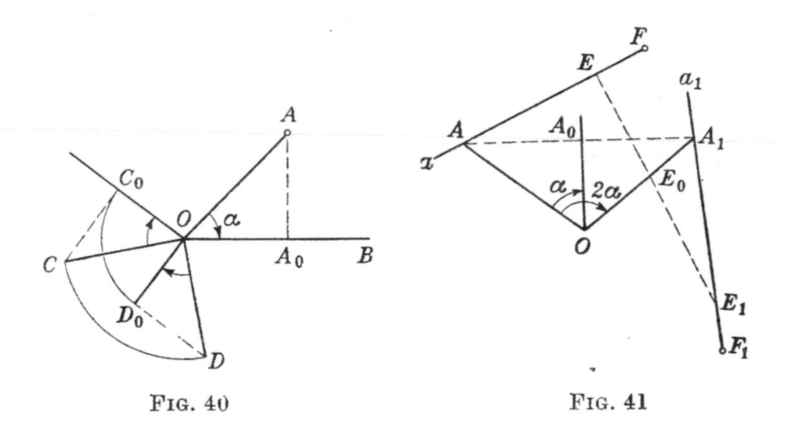

FIG. 40 FIG. 41

are congruent, whence the centres of the segments AA_1, EE_1, FF_1 lie on a straight line, by the theorem of Hjelmslev. But the centre A_0 of the segment AA_1 is precisely the point A_0 into which the transformation h maps the point A (since $\triangle AOA_1$ is isosceles, whence $\sphericalangle AOA_0 = \alpha$ and $AA_0 \perp OA_0$). The same applies to the points $E_0, F_0, \ldots$, and so the collinear points A, E, F have been mapped into the collinear points A_0, E_0, F_0.

It further follows quite simply that segments are mapped into segments. In Fig. 42a the arc AB with its chord have been mapped into the arc A_0B_0 with its chord.

A circle with radius OA and centre O is mapped into a circle with radius OA_0; a straight line which has one point in common with the former into a straight line with one point in common with the latter. Also the perpendicular to OA at A (the tangent to the former circle) will be mapped into the perpendicular to OA_0 at A_0 (the tangent to the latter). In other words *a right angle, one side of which passes through the centre O will be mapped into a right angle.*

Finally, we might easily show that transformation h is continuous—that is, if the points $A, B, C, \ldots$ tend to a point M, the corresponding points $A_0, B_0, C_0, \ldots$ will tend to M_0.

Summarising the properties of transformation h:

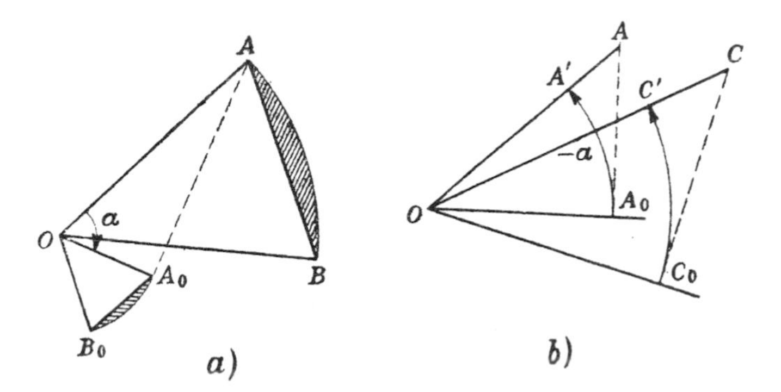

FIG. 42

It leaves the point O unchanged; maps segments into segments; angles with vertex O into equal angles with the same vertex; right angles with one side passing through O into similar right angles; circles with centre O into circles with centre O; and it is continuous.

It is somewhat inconvenient that the segment OA, when transformed into OA_0, not only becomes smaller but also rotates through the angle α. This, however, may be easily catered for: it will suffice, once the transformed figure has been obtained, to rotate it by the angle $-\alpha$ round O. It will then change only its position and point A_0 will be mapped into the point A' (Fig. 42b) and in effect point A will be transformed into the point

A' lying on the half-line OA, and similarly C into C', etc.

This final mapping, $A \to A'$, which we shall refer to as *mapping j*, obviously possesses all the above-enumerated properties of transformation h, from which it differs only slightly—namely, by the rotation round O.

Our transformations have been defined independently of the axiom of Euclid, and so they may be considered both in Euclidean and in non-Euclidean geometry. In the Euclidean they offer nothing new. In fact, (Fig. 42b),

$$OA' = OA_0 = OA \cos \alpha,$$

and we arrive at the point A' by laying off on OA a segment which is equal to OA multiplied by a constant number $\cos \alpha$; transformation j is simply a dilatation with respect to the centre O. We cannot use the formulae of ordinary trigonometry in non-Euclidean geometry, so we cannot assert that the ratio $\dfrac{OA'}{OA}$ has a constant value. We shall examine the consequences of this in the next section. Meanwhile we shall prove an absolute theorem which will be useful in § 10. This theorem is an immediate corollary of the properties of transformation h.

Let us take a quadrangle $OABB_0$ which contains two right angles A and B_0 (Fig. 43). Let us denote the angle BOB_0 by α and apply transformation h to the segment AB, taking O as the centre and as the angle of the transformation.

Point B will be mapped into B_0. To find the point corresponding to A let us construct the angle α on OA and drop the perpendicular AA_0 from A to its other side.

A_0 is the point we are looking for, the segment AB is mapped into A_0B_0, OA into OA_0. But the right angle OAB, one side of which passes through the centre O, is mapped, as we have learnt, into a right angle. Hence it follows that the angle OA_0B_0 obtained by transforming the angle OAB is right, and so A_0B_0 is the continuation of the segment AA_0; A_0B_0 and AA_0 form together a diagonal of the quadrangle. We now have the result:

Let two opposite angles of a quadrangle be right. Let us join their vertices and drop a perpendicular from one of the remaining vertices (from O in Fig. 43) to the diagonal thus obtained. Then, *this perpendicular forms with one side of the quadrangle passing through O an angle equal to that formed by the second side and the diagonal passing through O.*

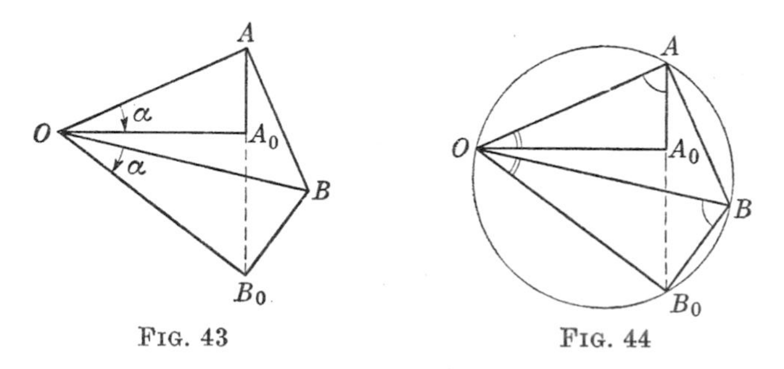

FIG. 43 FIG. 44

In Fig. 43 OA_0 is the perpendicular, OA and OB_0 are the sides passing through O, and the equal angles have been marked by α.

In Euclidean geometry this theorem follows easily from the properties of angles inscribed in a circle.

In fact, if the angles A and B_0 are right, it is possible to circumscribe a circle round the quadrangle (Fig. 44).

In the latter figure the angles marked with single arcs are equal because they stand on the same arc OB_0 and the angles marked with double arcs are equal because they represent the difference between 90° and the former angles (in the triangles OAA_0 and OBB_0). Q. E. D.

The theorem on inscribed angles does not belong to absolute geometry, so it is well worth noting that the theorem discussed above has been proved without appealing to it, in other words has been proved without recourse to the axiom of Euclid. Since this has been discovered only recently it testifies to the fact that in the field of elementary geometry, which seems to have been so thoroughly explored for hundreds of years, new discoveries are still possible.

§ 10. Mapping a plane into a circle

Let us return to the matters discussed in § 8 and recall that we deduced from the negation of the axiom of Euclid a corollary on the existence of an acute angle β, projections of points lying on one side of which onto the other side do not form a whole half-line but only a certain segment.

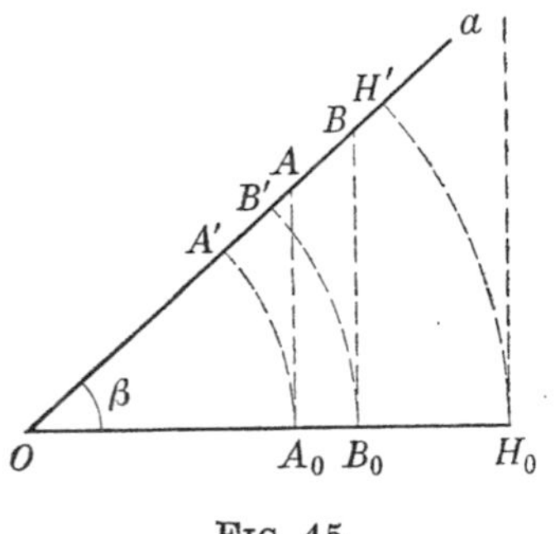

FIG. 45

We shall now make use of the result obtained in the last section: we will apply to the points of the Lobatchevskian plane the transformation j, choosing the centre O arbitrarily and taking the angle α of the transformation to be equal to β, spoken of above.

The projections of all the points of the side a (Fig. 45) onto the other side of the angle β will never fall outside the segment OH_0, although they may come as near as we like to its end H_0 (see § 8).

A rotation round O through $-\beta$ will map A_0 into A', B_0 into B'; the points A and B will be mapped by transformation j into A' and B'.

It is obvious that $A', B', \ldots$ will not go beyond the point H', where $OH' = OH_0$, but may come as close as we like to H'.

The infinite half-line OA is mapped into a finite segment OH', not including the point H'.

The point H' corresponds to no point of this half-line.

When A tends to infinity along the half-line, A' will tend to the point H'. We describe this by saying that the point H' represents an "infinitely distant" point H on the half-line a. On the half-line which is the continuation of a on the other side of O there is another infinitely distant point G which is mapped into G'.

What has been said of the half-line a may be repeated for any other half-line with origin O. Since all these half-lines are mapped into segments of length $OH' = r$, the entire plane is mapped into the interior of a circle with centre O and radius r. Each point of the interior of the circle represents a certain point of the plane. In our circle we have, so to speak, a map of the whole plane, a map which greatly diminishes distances in the neighbourhood of its perimeter, whose points correspond to the "infinitely distant" points of the plane.

This map reproduces segments as segments, so it gives infinite straight lines as chords of the circle.

Looking at our map we may become more easily acquainted with the properties of the Lobatchevskian plane. The points of the plane we shall always denote by capital letters $A, B, C, \ldots$. and the points corresponding to them on the map by the same letters supplied with a dash. Small letters will represent straight lines.

Let us consider the straight line a and the point C—shown on the map by the chord $D'E'$ and by C' respectively (Fig. 46).

Many chords pass through C'. Some of them, such as f', g', h', cut the chord $D'E'$, others, such as j', l', do not. The former, on the right of the picture, are separated from the latter by the chord $C'E'$ (e'). $C'D'$ (d') fulfils the same function on the left of the picture.

The chord e' represents a certain straight line e of the plane, and f', g', h', j' represent on the map the straight lines f, g, h, j (Fig. 46b).

The line e does not cut the line a (point E' lies on the perimeter of the map and does not correspond to any "finite" point; putting it differently, the lines e and a intersect at an infinitely distant point E which is represented on the map by E').

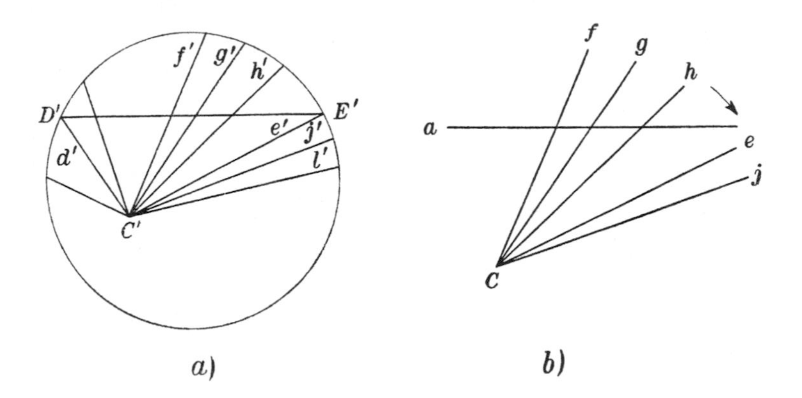

FIG. 46

If we rotated a movable straight line round the point C in the direction indicated by the small arrow in Fig. 46b, it would at first intersect the straight line a until it reached the position of e and thereafter not intersect it. This variable line would then occupy the position of the straight line j which also does not intersect a.

On the left-hand side of Fig. 46b the same thing is repeated, the straight line d ($C'D'$ on the map) now fulfilling the function of e. Joining both sides we obtain the figure shown in Fig. 47, which evidently possesses an axis of symmetry: the bisectrix of the angle *1* is at the same time perpendicular to *a*. There are two straight lines *e* and *d* passing through *C* which do not cut *a* (one of them passes through an infinitely distant point *E* on *a*, the other through the infinitely distant point *D*). Every straight line passing through *C* within the angle *1* will cut *a*, while none passing through *C* within the angles *2* and *3* will do so.

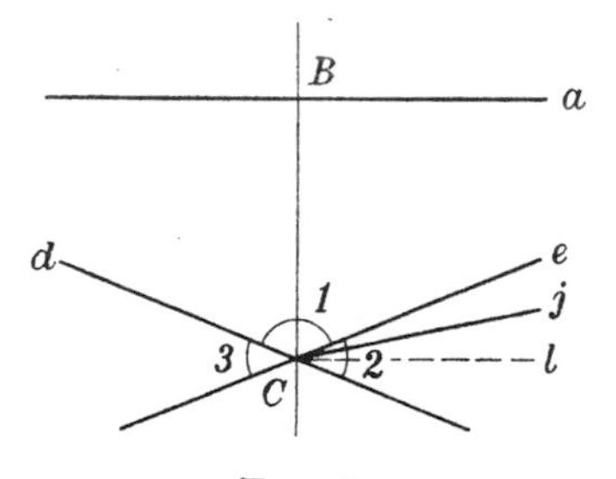

FIG. 47

Lobatchevsky called the lines *d* and *e parallel to a*, changing thereby the meaning of the term "parallel". Parallelism ceases to signify "not cutting"—for instance, the line *j* in Fig. 47 does not cut *a* either—but signifies "separating" the straight lines which cut *a* from those that do not. We shall sometimes, in order to attract notice to this change of meaning notation, supply the word "parallel" with the letter "(L)". Looking at our map again, we might also say that straight lines parallel to *a* and passing through *C* are those lines which connect *C* with the infinitely distant points on *a*.

Through a point not lying on a straight line there pass two straight lines which are parallel (L) *to it.*

One of them is parallel to *a* "on the right" (passing through the infinitely distant point *E*) and the other "on the left" (passing through the infinitely distant point *D*).

Straight lines which are parallel in the same direction are represented on the map by chords passing through one point of the perimeter and *vice versa*. It follows immediately that:

1. *If the straight lines a and b are parallel to the straight line c in the same direction they will be parallel to each other.*

2. *If* $a \| b$, *then* $b \| a$.

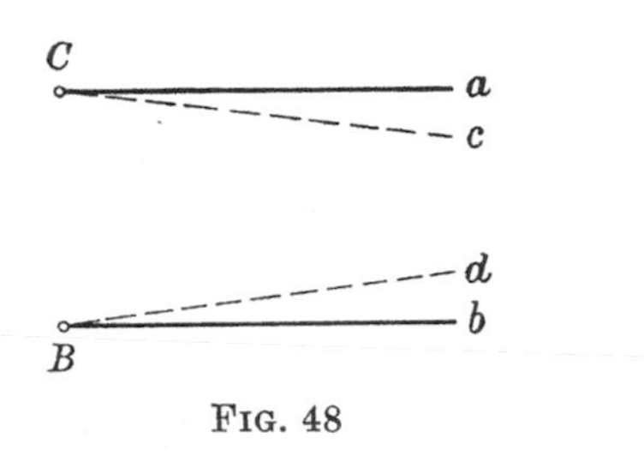

FIG. 48

The reader, perhaps, will shrug his shoulders and ask what the point could be of a sentence like the last one. Its gist, however, is quite clear in geometry: "$a \| b$" means (Fig. 48) not only that a and b do not intersect but also that close neighbours of a, e. g. c, do cut b. Similarly, "$b \| a$" requires that the line d, for instance, will cut a. Sentence 2 states that the above facts hold simultaneously.

Straight lines which do not intersect but which are not parallel are said to be *divergent*. Such straight lines are a and j in Fig. 47, and also a and the bisectrix l of the angle *2*. The lines a and l are both perpendiculars to CB, whence *two perpendiculars to the same straight line are divergent.*

Divergent straight lines are represented on the map by chords which have no point in common, e. g. $A'B'$ and $C'D'$ in Fig. 49. Let us connect B' and C' by the chord c', and A' and D' by d' and let these chords intersect at the point K'.

All the chords on the map (Fig. 49) represent straight lines in the Lobatchevskian plane; line c, and similarly line d, is in one direction parallel to a, and in the other to b.

The lines c and d pass through the point K, which on the map is represented by K' (Fig. 50). Thus on the plane we have a straight line a and two parallels to it passing through K, and also a straight line b with the same parallels.

If we recall the arguments concerned with Fig. 47 we see that the bisector of the angle *1* is the axis of symmetry of the whole figure and is perpendicular to a and b.

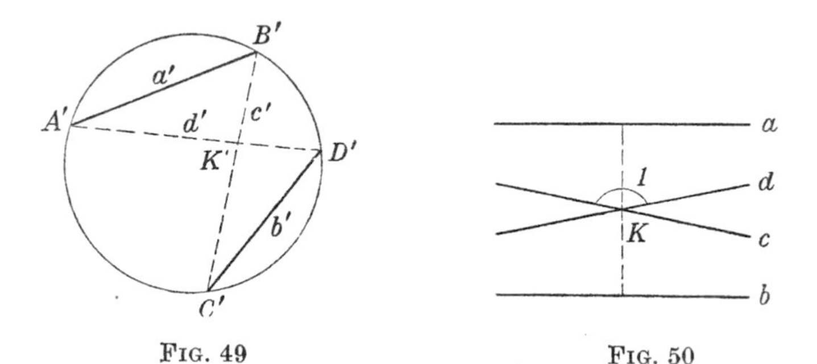

FIG. 49

FIG. 50

We have now arrived at the theorem that *there exist a common perpendicular to two divergent straight lines* We have stated above that two perpendiculars to the same straight line are divergent. Now we have proved the converse theorem.

Let us now summarize what we have been saying about the relative positions of two straight lines in a plane. They may

1° intersect, or

2° be parallel (that is, intersect at an infinitely distant point), or

3° be perpendicular to the same straight line.

All straight lines intersecting at one point form a pencil of Type I; straight lines passing through the same infinitely distant point, that is, straight lines parallel in the same direction, form a pencil of Type II; finally, all straight lines perpendicular to the same straight line, a pencil of Type III.

We have just been considering a straight line which is parallel to two divergent straight lines. Let us now take two intersecting straight lines which form an angle A with sides b and c and represent it on the map (Fig. 51) as the angle $B'A'C'$. The chord $a' = B'C'$ on the map

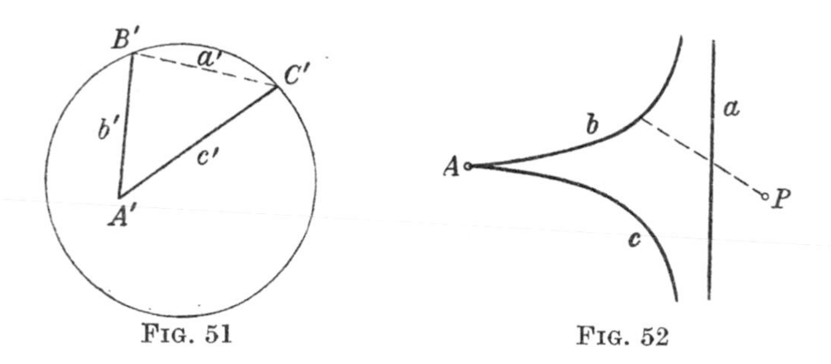

FIG. 51 FIG. 52

corresponds to a straight line which is parallel in one direction to side b and in the other to side c and which lies inside the angle.

In Fig. 52 we have shown the sides b and c as arcs of hyperbolae (cf. p. 61).

A straight line joining an arbitrary point of side b with a point P lying inside the angle and to the right of a cannot, it is clear, cut side c. Thus, a straight line passing through P can cut at most only one side of the angle. We face here a situation which appears paradoxical to all minds accustomed to Euclidean geometry. If we were to shade the angle by segments joining its sides we would cover only a part of the area between b and c, namely, that part which lies between b, c and a. Points to the right of a would be inaccessible. The reader who remembers the proof of Legendre quoted on p. 44 will now better understand why we are not entitled, unless we have assumed the axiom of Euclid, to say: "Let us construct, through a point inside an angle, a line to cut both its sides..."

There remains the case of the two parallel straight lines represented on the map by the chords b' and c' (Fig. 53).

We will not run through another argument similar to the last one, but we may say that *there exists only one straight line a which is parallel in one direction to the straight line b and in the other to the straight line c.*

Three such mutually parallel straight lines bound a certain area on the plane, an area extending to infinity since its boundaries are whole straight lines. Fig. 54 shows this area. It might be called a triangular area, or a triangle with infinitely distant vertices.

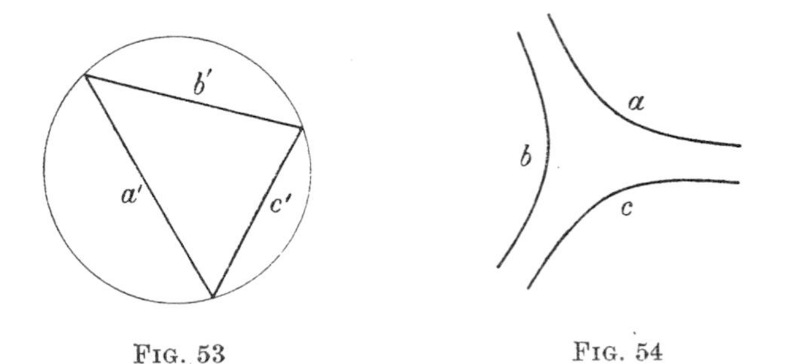

FIG. 53　　FIG. 54

We shall use this term; we shall also give the name "triangle" to an area bounded by the sides of an angle and a straight line parallel to both its sides, and to area contained between a segment and two half-lines connecting its ends with an infinitely distant point. In short, every area whose representation on the map is triangular will be considered a triangle.

A triangle, of whose vertices one, two or all three are infinitely distant points, will be known as an *improper triangle.*

The main instrument of research has been in this section the mapping of the Lobatchevskian plane into a circle. It may also be used with advantage in other problems, and the possibility of choosing quite arbitrarily the centre of this transformation furnishes the method with considerable flexibility.

1. As an example let us consider two non-parallel (intersecting or divergent) straight lines a and b. We will take the centre of the transformation O to be on one of them, say, on a. a has been represented on the map as the diameter a', b as the chord b' (Fig. 55) and straight lines perpendicular to a as chords perpendicular to the diameter a' because of the property of right angles one of whose sides passes through O.

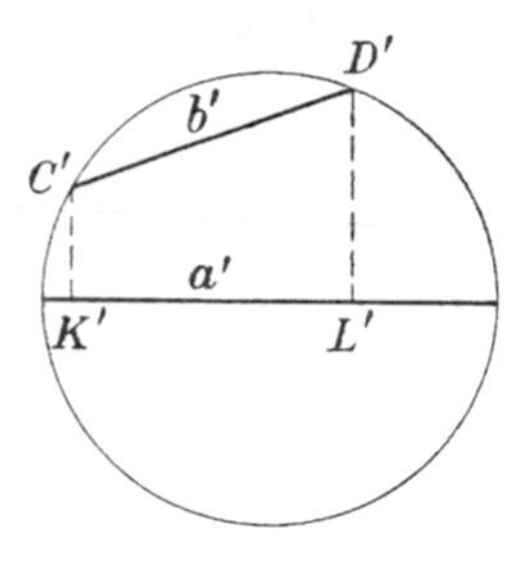

Fig. 55

The perpendicular projection of the chord b' ($C'D'$) onto the diameter a' is the segment $K'L'$, since by assumption neither C' nor D' lies on a'.

Returning from the map to the plane we see that the projection on a of every straight line b which is not parallel to a is a segment (in a particular case, a point).

In § 8 and at the beginning of this section we were concerned with this property for two straight lines intersecting at a certain angle β, the existence of which had been deduced from the negation of the axiom of Euclid. We now see that this reservation may be removed and the angle β may be replaced by any acute angle.

2. Another example: let us consider the improper triangle in Fig. 54. Choosing the centre of transformation on one of the three lines, say on a, we get the map in Fig. 56.

The perpendicular h' from A' to a' represents a certain straight line h parallel to both b and c, and perpendicular to a (Fig. 57).

The straight line b is the only one (cf. p. 76) parallel to a and h which lies inside the right angle *1*, and c is the only one parallel to a and h inside the angle *2*. It

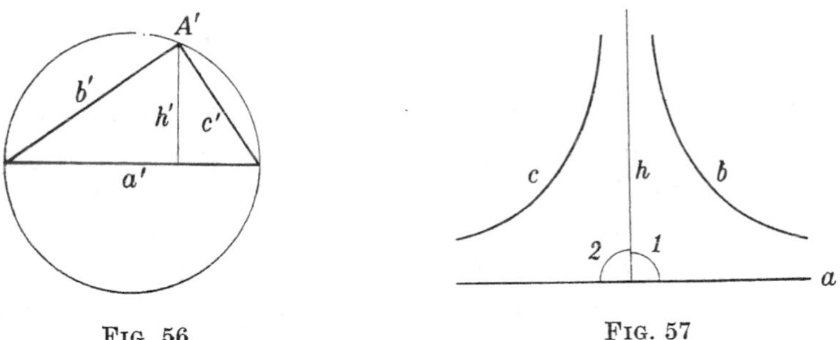

FIG. 56

FIG. 57

follows that the straight lines b and c are symmetrical about the axis h. Hence:

1° *A figure consisting of two parallel lines has an axis of symmetry parallel to them.*

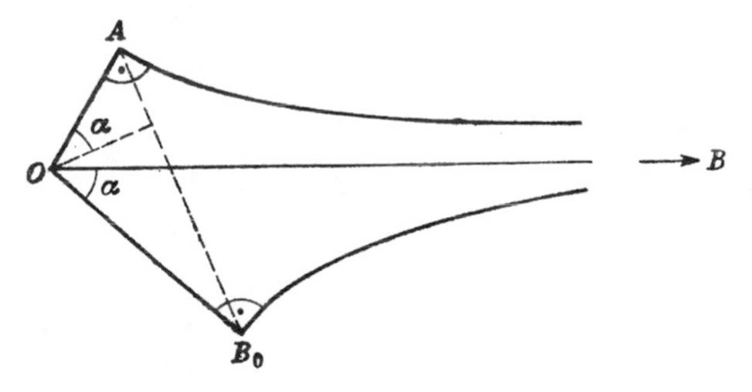

FIG. 58

2° *The improper triangle of Fig.* 54 *has three axes of symmetry.*

3. Finally, we shall show that our mapping enables us to extend certain theorems on polygons to improper

polygons. We assert that the theorem on p. 69 also holds if the point B is improper, as in Fig. 58. In fact, mapping this figure with centre O transforms it into an ordinary quadrangle altering neither the right angles A and B nor any of the angles with vertex at O. The theorem on p. 69 can then be applied to this ordinary quadrangle and it follows that the angles marked on Fig. 58 by α are equal.

§ 11. Angle of parallelism

In the last section we formulated the theorem that through a point not lying on a straight line a there pass two straight lines parallel (L) to it, and the bisector of the angle between the parallels is the axis of symmetry of the whole figure (Fig. 59). The acute angle α between this axis and one of the parallels is half the angle between these parallels and is known as the *angle of parallelism.*

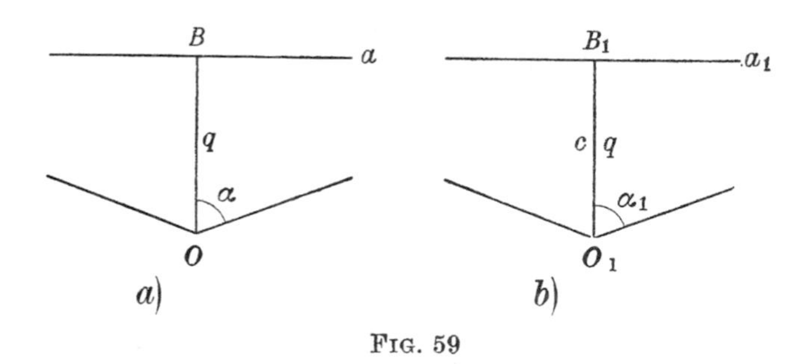

Fig. 59

Let us choose a fixed unit and express the length of the segment BO by q. If we chose the point O_1 at the same distance q from another line a_1 we would get the right-hand figure of Fig. 59, which may be placed to coincide with the left-hand figure, and so the angle of parallelism α_1 would be equal to α. This means that *the angle of parallelism depends only on the value of q, that is, that α is a function of q.* Lobatchevsky expressed this relationship:

$$\alpha = \Pi(q).$$

In order to learn something more about this function we shall make our map of the plane by choosing the point O in Fig. 59 as the centre for transformation j. The line OB becomes the diameter of the map and the line a becomes the chord a' perpendicular to this diameter, since right angles one side (here OB) of which passes through the centre O are reproduced on the map as right angles. The angle α is mapped into α' with vertex O (Fig. 60).

The sizes of angles whose vertex is at O remain unchanged on the map since their sides are unaltered, whence $\alpha' = \alpha$.

Let us move the chord a' perpendicular to OB' away from O.

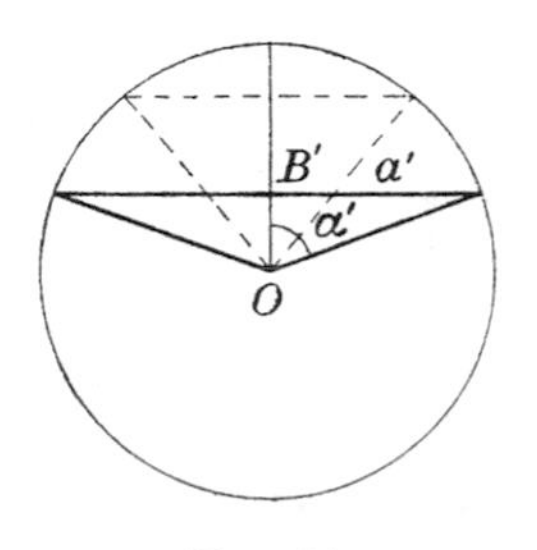

FIG. 60

It is evident from Fig. 60 that the angle α will then decrease. If the chord a' is sufficiently near to the centre of the circle the angle α will approach 90° and if the distance of the chord from O tends to the length of the radius of the circle the angle α will approach zero. Thus we get the corollary:

As q becomes larger the angle of parallelism becomes smaller, passing through all values between 90° *and* 0°.

In other words, as q goes from 0 to ∞ the angle of parallelism goes from 90° to 0°.

"Close to" the line a the angle of parallelism is practically right, and "far away" from it it does not differ much from zero.

"Close to" and "far away" are in inverted commas to stress the fact that they are indefinite terms. "Close to" on a sheet of paper means an inch or less, whereas when we are dealing with the structure of the universe the star α Centauri is considered to be near to the sun although it is something like thirty trillion miles away. Later we shall discuss in greater detail how is to be interpreted in objective reality that neighbourhood where the angle of parallelism is almost equal to 90°.

We have seen that the angle of parallelism α is a decreasing function of the variable q, but this is rather meagre information concerning the behaviour of the function Π. The problem naturally arises of how to determine this function more explicitly, so that we could "evaluate" the angle α when given the length q. The solution of this central and fundamental problem of non-Euclidean geometry was discovered by Lobatchevsky and Bolyai and testifies to their genius.

§ 12. Area of a polygon

First let us examine two parallel (L) lines cut by a third line.

We assert that two straight lines a and b which form with another line c equal corresponding angles 1 and 2 are perpendicular to a certain straight line h (Fig. 61). In fact, if we draw from the centre K of the segment AB the perpendicular KC to a and KD to b, we get congruent triangles KBD and KAC, so $\measuredangle 3 = \measuredangle 4$ and KC and KD lie on the same straight line.

Thus *straight lines which form with another straight line equal corresponding angles are divergent.*

The straight line d passing through B and parallel on the right to a must go "below" line b, which is divergent with a, so the angle 5 formed by c and d will be greater than the angle 1, and consequently also than 2.

In order to put this in a simpler manner let us bear

in mind that the angle *2* is an interior angle of the improper triangle whose sides are d, BA and a, while *5* is an exterior angle.

Hence, *the theorem on the exterior angle applies also in the case of an improper triangle.*

Now let us turn to the theorem on the sum of the angles of a triangle.

In chapter I we considered this topic in detail when following the historical development of the theory and we proved that in the geometry which is based on the

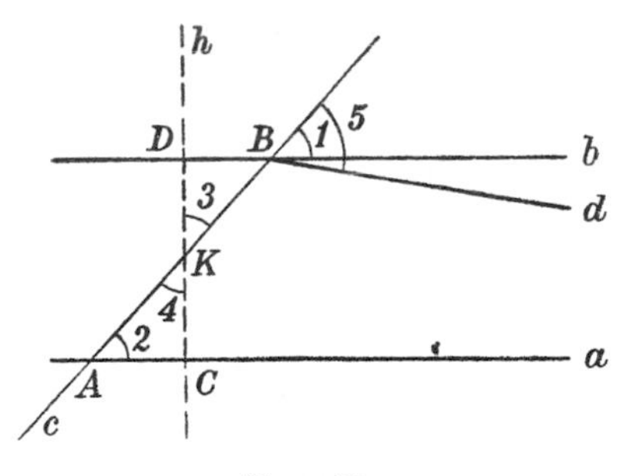

FIG. 61

negation of the axiom of Euclid the sum of the angles of a triangle is less than 180°. However, since in this chapter we wish to give an exposition of non-Euclidean geometry without using the results of chapter I, we shall work it out again from scratch. At the same time we shall examine some figures and constructions which are interesting in themselves and very characteristic of Lobatchevskian geometry.

Perhaps the most basic figure of the Euclidean planimetry in the rectangle. We shall see that in non-Euclidean geometry the sum of the angles of a quadrangle is less than 360°, so that rectangles cannot exist. The simplest quadrangle is the *quadrangle of Saccheri*. Saccheri was the first to observe the relationship between the properties of this quadrangle and the negation of the axiom of Euclid.

By the quadrangle of Saccheri we shall understand the quadrangle $ABCD$ (Fig. 62) in which angles A and B are right and the sides next to them, BC and AD, are equal.

This quadrangle has an axis of symmetry h which is perpendicular to AB at its centre.

The points C and D are symmetrical about h, so the segment CD is perpendicular to h.

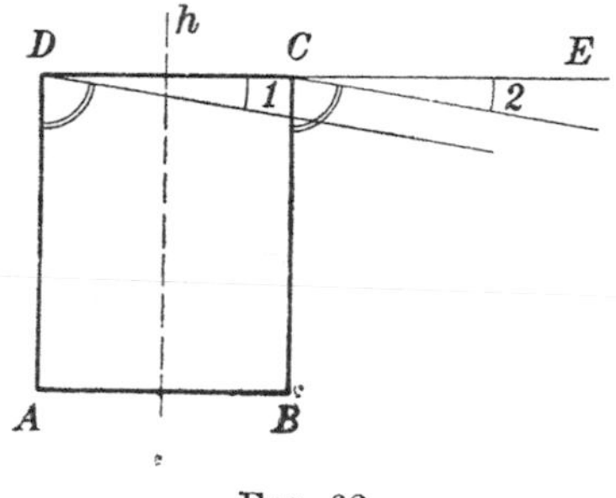

FIG. 62

The axis of symmetry divides the quadrangle of Saccheri into two quadrangles each of which contains three right angles.

We shall show that *the sum of the angles of the quadrangle of Saccheri is less than* 360°. It will suffice to prove that the angles C and D are acute.

Let us draw through C and D two straight lines which are parallel on the right to AB; together with the segment CD they form an improper triangle, so the exterior angle *2* is greater than the interior *1*. On the other hand, the angles marked in the figure by two small arcs are equal, since they are the angles of parallelism corresponding to equal distances from AB. Therefore

$$\sphericalangle ECB > \sphericalangle CDA.$$

But $\sphericalangle CDA = \sphericalangle DCB$, since they are symmetrical about h, so

$$\sphericalangle ECB > \sphericalangle DCB.$$

These two angles add up to 180°, so that angle DCB, the angle of our quadrangle, is acute. Q. E. D.

In the quadrangle of Saccheri two angles are right and two are acute.

The sum of the angles of a triangle is less than 180°.

In our proof we shall use the construction known in Euclidean geometry as the transformation of a triangle into an equivalent rectangle.

Let the angles A and B of triangle ABC be acute. Join the centres of the sides AC and BC by a straight line m (Fig. 63) and drop perpendiculars to this straight line from all the vertices of the triangle.

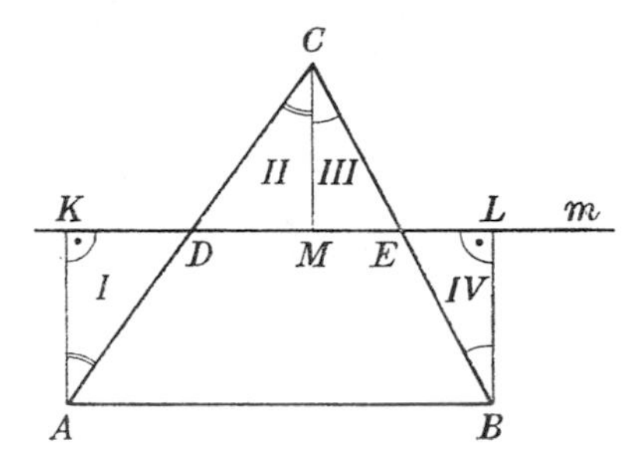

FIG. 63

It is easy to prove (as in Euclidean geometry) that the triangles I and II, as also III and IV, are congruent. It follows that the angles marked in Fig. 63 by single or double arcs respectively are equal and that

$$AK = CM = BL.$$

The sum of the angles of the triangle consists of the angles A and B and of the two angles marked by a single and a double arc, and is therefore equal to the sum of the angles KAB and LBA.

But the quadrangle $KLBA$ is a quadrangle of Saccheri with base KL and right angles K and L. Therefore the angles KAB and LBA are acute and their sum, which is equal to the sum of the angles of the triangle ABC, is less than 180°. Q. E. D.

The above construction divides the triangle into three parts: the quadrangle $ADEB$ and the triangles II and III. By arranging the triangles differently these parts form together the quadrangle of Saccheri. For this reason we may refer to the construction as the replacement of a triangle by an equivalent quadrangle of Saccheri.

It follows from the theorem on the sum of the angles of a triangle that the sum of the angles of a quadrangle is less than 2.180° and that the sum of the angles of a polygon of n vertices is less than $(n-2).180^\circ$.

Definition. By the *defect of a polygon* of n sides we understand the difference between $(n-2).180^\circ$ and the sum of its angles; in other words, the defect of triangle ABC is $180^\circ-(\sphericalangle A+\sphericalangle B+\sphericalangle C)$.

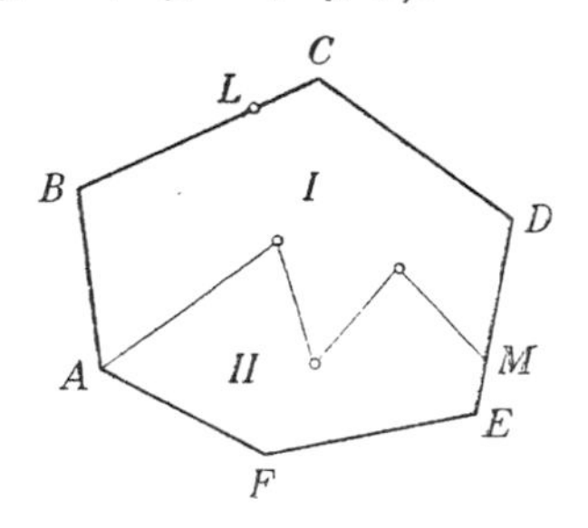

Fig. 64

Thus it follows that the defect of any polygon is a positive number.

The concept of the defect of a polygon is very important in non-Euclidean geometry, as is shown by the following theorem:

The defect of a polygon divided into two polygons by a polygonal line is equal to the sum of the defects of the component polygons.

Proof. First of all let us note that we are entitled to adjoin to the vertices of the polygon (Fig. 64), for instance, the point M, without altering its defect. In fact, $(n-2).180^\circ$ becomes larger by 180° (since n increases by 1), but the sum of the angles of the polygon increases by the angle at M, which is in fact 180°.

Therefore we may assume that the polygonal line which divides the polygon begins and ends at the vertices of the latter. We have

$$\text{Defect of part } I + \text{Defect of part } II = (p_1-2+p_2-2).180^\circ - \alpha,$$

where p_1 and p_2 denote the numbers of the vertices of the component polygons and α the sum of all their angles.

Let r stand for the number of vertices of the polygonal line which lie within the polygon, β for the sum of the angles of the whole polygon and n for the number of its vertices.

We have

$$p_1+p_2 = n+2+2r,$$
$$\alpha = \beta + r.360^\circ,$$

and so

$$\text{Defect}\, I + \text{Defect}\, II = (n+2r-2).180^\circ - \beta - r.360^\circ$$
$$= (n-2).180^\circ - \beta,$$

i. e. it is equal to the defect of the whole polygon. Q. E. D.

The concept of defect may be expanded to cover improper polygons, for example the triangle ABS (Fig. 65), in which S is an infinitely distant point. However, we must first define the size of the "angle" S. Of course, the definitions of the theory of measuring angles cannot be applied here, since they rest on taking a chosen unit, for instance, one degree, and comparing it with the angle under consideration. This process breaks down here, and we are forced to bring in the measure of the "angle" S independently. We assume the magnitude of angle S

to be 0, and so the defect of triangle ABS will be $180° - \sphericalangle A - \sphericalangle B$.

Such an extension of the concept of defect make sense only if the theorem on the defect of a polygon consisting of parts applies also to improper polygons.

This is indeed the case. Without going into a full, general proof we may content ourselves with the case in Fig. 65:

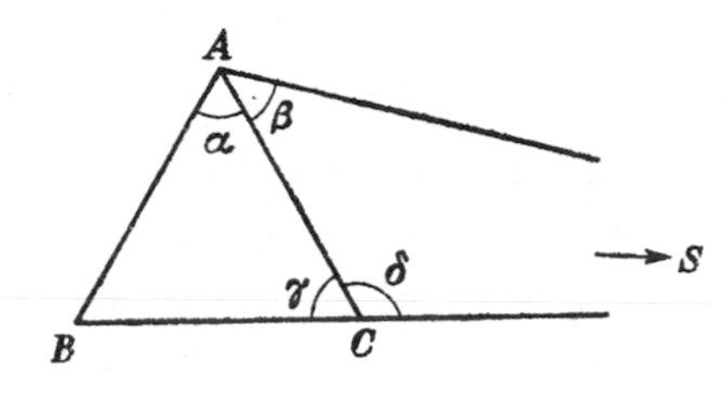

FIG. 65

$$\text{Defect } \triangle ABC = 180° - \alpha - \sphericalangle B - \gamma,$$

$$\begin{aligned}
\text{Defect } \triangle ACS &= 180° - \beta - \delta, \\
\text{Defect } \triangle ABC + \text{Defect } \triangle ACS &= 360° - (\gamma + \delta) - \\
&\quad - \alpha - \beta - \sphericalangle B \\
&= 180° - \alpha - \beta - \sphericalangle B \\
&= \text{Defect } \triangle ABS.
\end{aligned}$$

According to the definitions adopted the defect of the triangle in Fig. 54 is 180°. Obviously no triangle may have a greater defect. Similarly, no quadrangle may have a defect greater than 360°.

We said on p. 83 that a rectangle does not exist, and so a square cannot exist either in Lobatchevskian geometry. A plane cannot thus be latticed, and a square foot divided into 144 square inches cannot exist. Thus the measuring of areas by means of square feet and inches must of necessity break down. Does this mean that we should give up the concept of the area of a polygon? Obviously it does not. Triangle ABC and quadrangle $AKLB$ in Fig.

63 consist respectively of the same parts, so why should we hesitate to consider their "sizes", that is, their areas, to be equal? The areas of polygons are quantities also in non-Euclidean geometry, and we must only define the concept of area with sufficient skill in order that it may be as useful in Lobatchevskian geometry as was in Euclidean geometry the definition based on the grid of squares.

Which property of area is of the greatest practical importance? Obviously that which states that a three-acre field together with a two-acre field give a five-acre plot of land; more generally, that the area of a figure consisting of two parts is equal to the sum of the areas of the parts.

In non-Euclidean geometry this property (known as *additivity*) is possessed by the defect of a polygon and also by the product of a defect and a constant number. So we shall satisfy all practical requirements if we adopt the following definition:

By the area of a polygon we understand the product of its defect and a constant λ. The value of this constant depends on which polygon has been chosen to have unit area, that is, it depends upon the unit.

Therefore the area of a triangle with angles A, B, C is

$$[180° - (\sphericalangle A + \sphericalangle B + \sphericalangle C)].\lambda .$$

Research, which we cannot go into here, has explained that the concept of the area of a polygon is connected with the division of the polygon into parts. If the areas of two polygons calculated according to the above definition are equal, one of them may be cut up into parts out of which the other may be constructed, just as the quadrangle of Saccheri in Fig. 63 was made up of parts of triangle ABC. In other words, polygons with equal defects consist of correspondingly equal parts. This property forces us to consider such polygons as being "equal in size"

The reader may entertain the following doubt: in sufficiently small domains, like those dealt with in chapter I, the angle of parallelism is practically 90° (we shall come back to this later on) and Euclidean geometry is valid with a great degree of precision; in particular the area of a triangle is half the product of its base and its altitude. Is this consistent with the definition that the area of a triangle is equal to its defect times a constant λ?

There is no inconsistency whatever. The defects of "small" triangles are close to zero, but the constant λ is so large that the product of the defect and λ ceases to be a tiny quantity and assumes approximately the value of half base times altitude. Later we shall approach this question again and at the same time examine the constant λ.

§ 13. Regular polygons

In Euclidean geometry, if a polygonal line has equal sides and equal angles, as in Fig. 66a, its vertices will lie on a circle. This is not so in non-Euclidean geometry. Let, say, c and d be two parallel lines and the point K be equidistant from them.

We find points L and M symmetrical to K about both straight lines and construct a polygonal line with equal sides and equal angles and having LK and KM as its initial sides. It is obvious that the vertices of the polygonal line thus obtained cannot lie on a circle because its centre should be the point of intersection of c and d—whereas these lines do not intersect. We see that *there is no circle* passing through the points K, L, M (Fig. 66b). Continuing the construction we get an infinite number of parallel lines like $c, d, e, \ldots$ and an infinite number of sides of the polygonal line. This polygonal line could be called a regular one with an infinite number of sides. It would undoubtedly be a figure of great interest, but here we shall confine ourselves to the examination only of regular polygons in the usual sense of the term. We shall adopt the definition:

A *regular polygon* is a polygon inscribed in a circle with equal sides.

The angles of this polygon will also be equal.

The easiest to construct is a *regular quadrangle* (*a square*). Let us take segments $OA = OB = OC = OD = r$ (Fig. 67) on the axes of a perpendicular system of coordinates

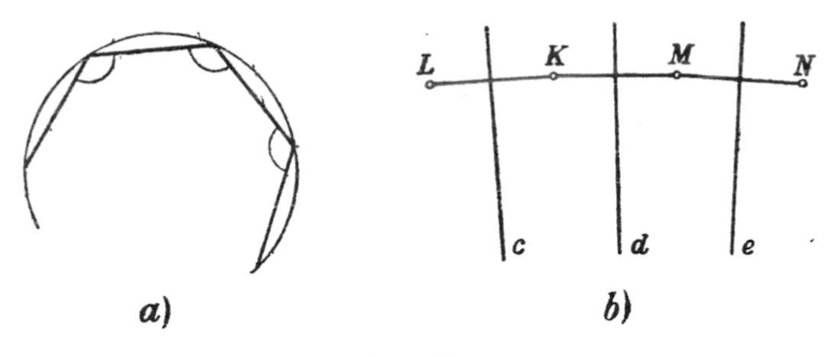

FIG. 66

and join their ends. We obtain a regular quadrangle with angles 2α. Since the sum of the angles of a quadrangle is less than 360° we have $2\alpha < 90°$.

When r increases the new quadrangle $A_1B_1C_1D_1$ will contain the previous one, and therefore its defect (and hence its area) will be greater than that of the first quadrangle and the angle α_1 will be smaller than the angle α.

The greater the square, the smaller its angles.

Draw through vertex B a line b parallel to OA. The angle of parallelism β is obviously greater than α, so

$$2\alpha < 2\beta.$$

If the segment r tends to infinity, the angle of parallelism β tends to zero (p. 80), so 2α also tends to zero. Therefore squares may exist whose angles may be as small as we like—for instance a square with angles of one degree. A strange square this, to our minds enslaved by Euclid! One might compare it with the curvilinear figure of Euclidean geometry given in Fig. 67b.

On the other hand the defect of the triangle OAB is less than the defect of the improper triangle OBS containing it, so we have the inequality

$$180° - (90° + 2\alpha) < 180° - (90° + \beta),$$

and further

$$2\alpha > \beta.$$

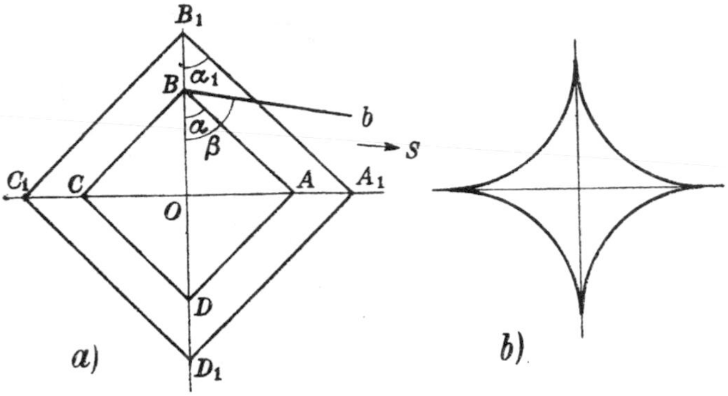

FIG. 67

If the segment r tends to zero, the angle of parallelism β will tend to $90°$, so also, therefore, the acute angle 2α. Sufficiently small squares have angles approximating to $90°$.

The shapes of squares in non-Euclidean geometry are various; their angles may assume all values between $90°$ and $0°$.

Since the angle of a square is less than $90°$, four squares with a common vertex O do not fill the full angle $360°$ at this vertex, and therefore, as stated above, a plane cannot be latticed—that is, composed of squares any four of which would have a common vertex. This does not, however, mean that it is impossible to fill the Lobatchevskian plane without gaps by means of squares. There exists, for instance, a square with angle 60°. Let us put six of these squares together so that they have a common vertex O and do not overlap. The angles at O will add up to 360°, and no gap will appear between the squares. If we repeat the same process at the outer vertices of the squares, that is, if we supply more squares so that there will be six meeting at any one vertex, we shall cover the entire plane by squares.

For such "parquet-work" we could also use squares with angle 45°, 72° or more generally $\frac{360°}{n}$ (when $n > 4$).

It is a very interesting fact that these arguments, which look like a mathematical amusement, have proved important in the theory of functions. The fates of mathematical theories are impossible to foresee.

The construction of an *equilateral triangle* in non-Euclidean geometry is similar to that in Euclidean: however, the angle of such a triangle is not 60° but less. All the

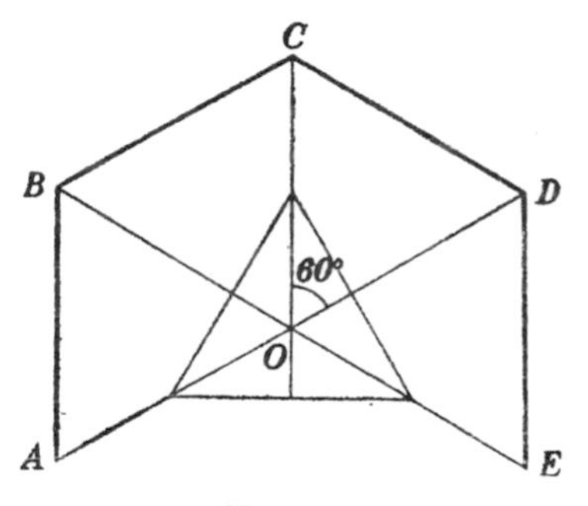

FIG. 68

same, one may use an equilateral triangle to construct an angle of 60°; it is necessary to connect the centre of the triangle (Fig. 68) with two of its vertices and then to bisect the angle thus obtained.

A *regular hexagon* may be constructed by marking off equal segments $OA = OB = OC = \dots = r$ on the sides of consecutive angles of 60° (Fig. 68).

One should not think of a regular hexagon as consisting of six equilateral triangles. This is not so, since angle OBC is less than angle BOC.

The reader may easily verify that the angle of a hexagon decreases as r increases; this angle will be 90° for a certain value of r. This offers us the possibility of constructing a parquet out of regular hexagons any four of which meet at one vertex.

We see that the theory of regular polygons in non-Euclidean geometry is incomparably richer in problems than that in Euclidean.

§ 14. Divergent and parallel straight lines

We have shown in § 10 that the perpendicular projection of one of two divergent straight lines onto the other, e. g. the projection of c onto d in Fig. 69a, is a segment.

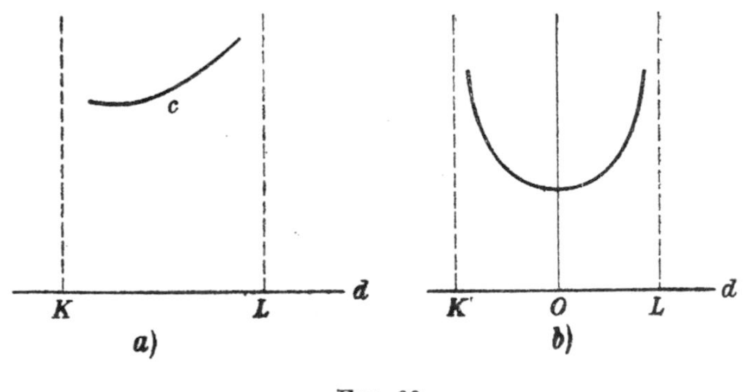

FIG. 69

It follows from this that c is situated between two perpendiculars to d at the ends of this segment.

Straight lines on the Euclidean plane do not possess this property, but it is possessed by, for instance, the curve in Fig. 69b, which is the graph of $y = \dfrac{1}{\sqrt{1-x^2}}$.

The perpendiculars drawn dotted in the figure are the

asymptotes to the curve and its projection onto d is a segment. If we travel along this line at a constant speed we move infinitely far away from the horizontal axis. Similarly, the straight line c, on the non-Euclidean plane, also departs infinitely far from d, and, what is more, in both directions.

We shall now prove that *the shortest distance between two divergent straight lines is their common perpendicular* (i. e. the segment perpendicular to both lines).

First, a simple observation. Let $ABCD$ be a quadrangle of Saccheri (Fig. 70). Produce AD to E. The angle E will be smaller than angle ADC by virtue of the theorem on the exterior angle of a triangle.

The angle BCE is greater than the angle BCD which is equal to the angle ADC.

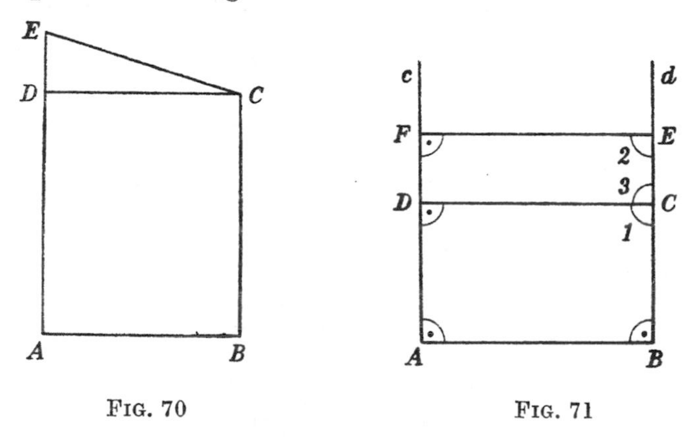

FIG. 70 FIG. 71

Finally: *if in a quadrangle $ABCE$, where A and B are right angles, side AE is longer than the side BC, then that upper angle is smaller which lies beside the longer side.*

The converse theorem is obviously true also. We shall make use of it when considering the straight lines c and d, divergent from each other and perpendicular to AB (Fig. 71).

Let us drop the perpendiculars CD and EF from C and E to the line c. The angle *1* is acute, since the quadrangle $ABCD$ already has three right angles. Similarly, angle *2* is acute. Angle *3*, adjacent to an acute angle, is obtuse, and so greater than angle *2*. We apply the above lemma now to the quadrangle $FDCE$, in which angles D and F are right. We obtain the inequality

$$EF > CD.$$

This signifies that as we travel upwards along d we get further and further away from c, having been closest to it at B. Q. E. D.

Of course, the same sort of thing can be said when we travel downwards along d.

We shall now quote without proof a certain property of divergent straight lines which will be proved on p. 159 and which is bound up with the segment perpendicular to both divergent lines.

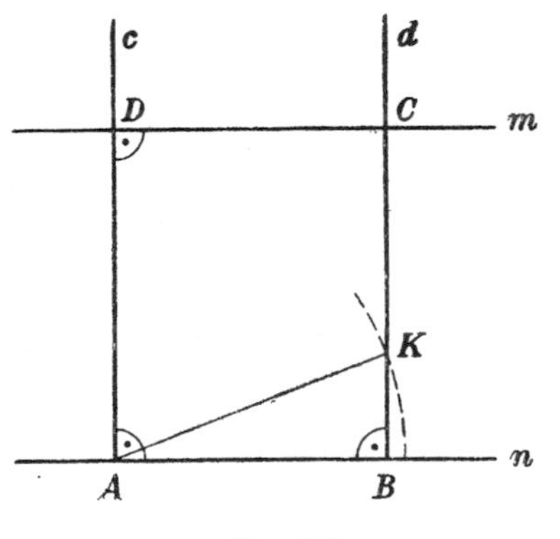

FIG. 72

Let c and d (Fig. 72) be divergent straight lines and let AB be the shortest distance between them.

Drop the perpendicular CD to c from a point C chosen on d.

As we know

$$CD > AB.$$

Let us construct a circle with centre A and radius CD. Owing to the inequality above it will intersect d at the point K.

The property we mentioned states that *the straight line AK is parallel to CD.*

Bolyai made this by no means simple discovery and used it to show that a line parallel to a given line may be constructed, by means of a ruler and a pair of compasses.

The construction is as follows:

In order to find the line parallel to m through A (Fig. 72) we first drop the perpendicular AD to m from A, and then the perpendicular n at A to the line AD just obtained.

We then choose on m an arbitrary point C (different from D), and drop from it a perpendicular to n.

So far our construction has reproduced the figure of the theorem of Bolyai, all but the point K. This is found as above by drawing the circle with centre A and radius CD.

AK is the parallel line we wanted.

Naturally, the value of this construction is theoretical, since it assumes that we have at our disposal a pair of compasses with legs of any length.

As an exercise we shall leave to the reader the proof of the following theorem: *if we travel along one of two parallels c and d in the direction of their parallelism we shall approach ever more the other* (Fig. 73). It is similar to the proof of Fig. 71.

The reader should here be warned not to progress too readily to the corollary that as we travel towards infinity along c on the right-hand side the distances of points on c from the line d will tend to zero. For instance, the function $1+\frac{1}{x}$ $(x > 0)$ decreases as x increases, but its value is always greater than one, and cannot be said to tend to zero.

Neither would an appeal to the map, on which the two parallels are represented by diameter c' and chord d' starting from P' (Fig. 74), be convincing. True, the segment $K'L'$ tends to zero as K' tends to P', but it does

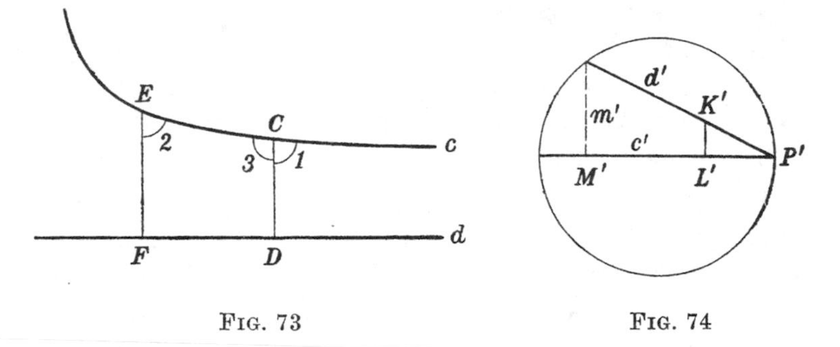

FIG. 73

FIG. 74

not mean that the corresponding segment KL on the Lobatchevskian plane will do so. The shortness of $K'L'$ might be due to the fact that near the edge of the map all segments are shortened considerably.

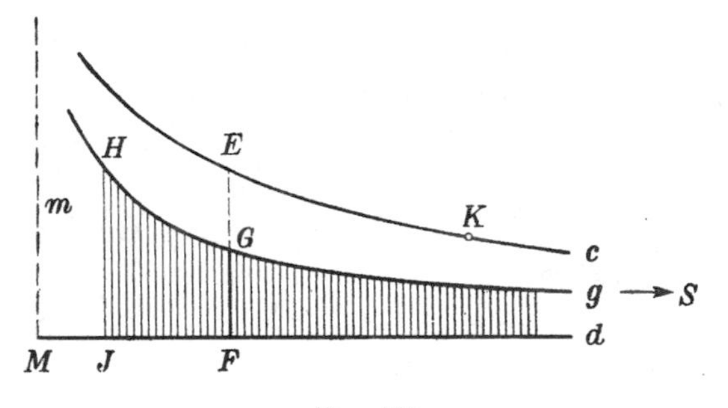

FIG. 75

In order to prove that parallel straight lines approach asymptotically to each other, i. e. that the distance of C from d (Fig. 73) tends to zero as C approaches infinity (on the right-hand side) along c, let us turn to another argument.

First let us observe that the perpendicular projection of c onto d is the half-line MS (see Figs. 74 and 75), so that when we travel to the left along c we get further and further away from d. Let us now choose any small segment ε, take $FG = \varepsilon$ and draw through G a parallel g to d (on the right). This line rises infinitely high on the left, and there is a certain point on it which is as high above the axis d as is E. Let us call this point H. $HJ = EF$.

If we now shift the shaded part of the figure along d to the right through a distance equal to JF, H will fall on E and the straight line g, which is parallel to d, will coincide with that parallel to d which passes through E, i. e. with c. At the same time the point G will fall on a certain point K. All the points of c lying to the right of K are distant from d by less than ε. Consequently c comes infinitely close to d. Q. E. D.

§ 15. Circular lines

The title of this section may be surprising. "Why "circular lines" and not "circles"?" some readers may ask. Surely the geometrical locus of points equidistant from a fixed point is a concept which remains unchanged even in non-Euclidean geometry. It is in fact so with this geometrical locus, but let us reflect whether this is the only characteristic of a circle.

We might say yet that a circle is a finite line which can move along itself.

Perhaps the latter property plays a greater rôle in practice than the former. An infinite line with this property is a straight line. No other such lines exist on the Euclidean plane; in space there appears in addition only the circular helix.

Further, the circle is the only finite line which may be divided in half in an infinite number of ways, that is, which has an infinite number of axes of symmetry. A straight line also has an infinite number of axes of symmetry: all the straight lines perpendicular to it.

Both the properties quoted, the former which refers to the motion of a circle, or a straight line, along itself and the latter which refers to symmetry, are rather primary ones. It may easily be shown that the former follows from the latter.

It is not essential in Euclidean geometry whether we choose one of them as the basis of our study of circle or the ordinary definition which gives the circle as a set of points equidistant from the centre. Since both these definitions are equivalent. However, in non-Euclidean geometry, it turns out that the properties dealing with this motion and with symmetry are possessed not only by straight lines and circles but also by other lines which are called equidistants and horocycles.

The reason for this is that the properties of the set of all possible motions and symmetries are different in non-Euclidean geometry from those in Euclidean.

In both geometries one may shift a figure along a certain straight line m; in Fig. 76, for instance, the triangle ABC may be moved into the position $A_1B_1C_1$.

In Euclidean geometry the segment CC_1 is equal to AA_1, or the distance travelled by the apex of the triangle is equal to that travelled by one of the lower vertices.

This is not the case in non-Euclidean geometry: the lines BC and B_1C_1 are perpendicular to m and the shortest distance between them is BB_1, so that

$$CC_1 > BB_1.$$

The apex C travels further than B; the head of a walking man moves more quickly than his feet—just as it happens on the earth when we walk along the equator, since the head is then more remote from the centre of the earth than the feet.

Furthermore, the line traced by the point C when the triangle ABC continually changes its position with side AB sliding along m is by no means a straight line.

In fact, BCC_1B_1 is a quadrangle of Saccheri. The segment DE of the axis of symmetry of the quadrangle is perpendicular (p. 84) to the sides BB_1 and CC_1, whence it follows that it is the shortest distance between these sides. That is,

$$DE < BC.$$

This means that when the triangle moves so that B reaches the position D the apex of the triangle will be above the segment CC_1, and therefore C traces a curve with chord CC_1 above CC_1, i. e. it is convex upwards.

A line consisting of points equidistant from the straight line m and lying on the same side of m will be referred to as an *equidistant* ([1]).

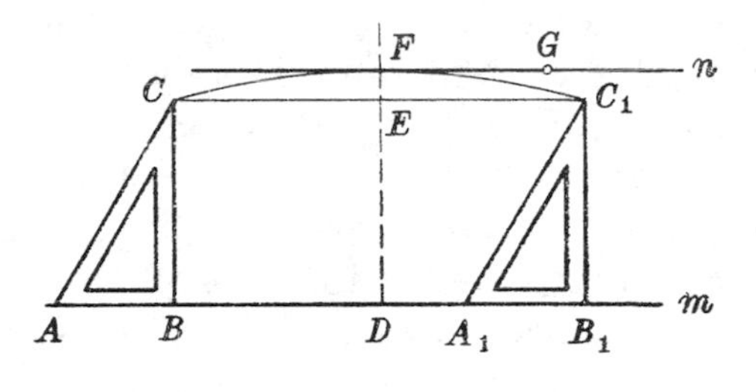

FIG. 76

The line m is known as the *base* of the equidistant and the distance of the points on the line from the base as the *altitude* of the equidistant.

When in the moving triangle ABC one side of it slides on the straight line m, each point of the triangle which does not lie on this side traces an equidistant and each point of this side traces the straight line m, which might be considered as an equidistant whose altitude is zero.

Equidistants with the same altitude may be placed to coincide, so that they are congruent figures.

Let us now consider the domain between the equidistant and its base m (Fig. 76) and let us shift it so that m slides along itself. Every point on the equidistant remains at

([1]) Some authors call it a hypercycle.

a constant distance from m, so therefore the equidistant moves along itself and any point on it may assume the position of any other one. Consequently, an equidistant possesses the first of the properties of the circle which were discussed at the beginning of this section.

In non-Euclidean geometry tram-lines could not be straight; if one of them were straight the other would be an arc of an equidistant. Similarly, the sides of a drawer would have to be arcs of equidistants, for otherwise the drawer could not be smoothly pulled out.

Let us now consider the axes of symmetry of an equidistant.

All straight lines perpendicular to the base of the equidistant, e. g., DE in Fig. 76, therefore all lines of a pencil of type III (p. 76) are axes of symmetry. Let the axis of symmetry of an equidistant cut it at F.

The straight line n perpendicular to the axis of symmetry and passing through point F is divergent from the base m. The shortest distance between these lines is the segment DF, so that any other point G of n lies further from m than F, lies, in other words, above the equidistant. The straight line n has only one point F in common with the equidistant; we shall call it a *tangent* to the equidistant. The direction of the tangent to a line at a given point is known as the *direction of the line* at this point. Therefore, the direction of the equidistant at its point F is perpendicular to FD, or, briefly, *an equidistant cuts at right angles all straight lines perpendicular to its base* (straight lines of a pencil of type III), just as a circle cuts at right angle all lines passing through its centre (straight lines of a pencil of type I).

The study of the equidistant is connected with that of shifts along a given straight line m. All such shifts form, as we say, a group, which means that if we carry out two successive shifts along m we will have the same result as if we had carried out one certain shift. Shifting, for example, the segment BC into the position DF (Fig. 76), and then DF into the position B_1C_1 has the same effect as shifting BC along m directly to the position B_1C_1. We should add that there is always an "opposite" shift to any given one; BC may be shifted into the position DF, but conversely one may also shift DF into the position BC.

In Euclidean geometry a group is formed not only by these shifts, or translations, along the same straight line; all possible translations also form a group. The quadrangle in Fig. 77 has been translated first into position *II* and then into position *III*.

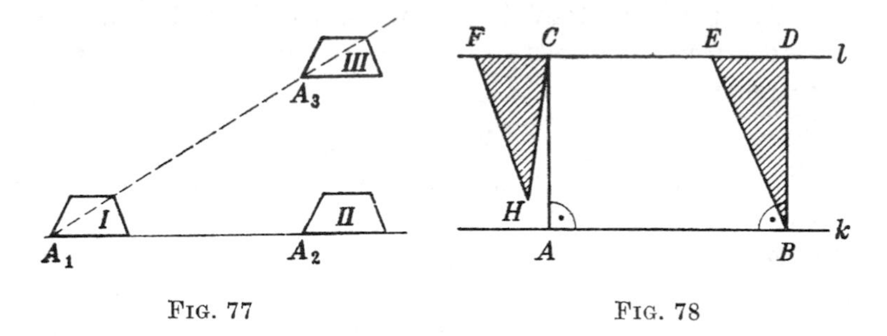

FIG. 77

FIG. 78

The same result could be obtained by translating the quadrangle along the straight line A_1A_3. The successive carrying out of two translations is equivalent to the one resultant translation. This fact is better known in the following form: the resultant of vectors A_1A_2 and A_2A_3 is the vector A_1A_3. We will adhere, however, to our original wording, which states that if we combine a number of translations the result will be again a translation.

This is not true for non-Euclidean translations, as we shall show by an example (Fig. 78).

Let a certain translation along the straight line k give AC ($\perp k$) the position BD.

Let us then translate the segment BD along the straight line l, which joins C and D, so that D takes up the position C.

Let us imagine, for example, that we fix a cross-bar DE to BD and translate it along l until it assumes the position CF. Angles BDE and ACD are upper angles of the quadrangle of Saccheri $ABCD$, so they are acute (p. 85), while the angle ACF is obtuse. Consequently the side BD of the shaded acute angle will not coincide with CA but lie within the angle FCA—that is, in the position CH.

We now see that as the result of the two translations the point C has returned to its initial position but the point A has not. The same final result would have been obtained by rotating the segment CA round C through a suitable angle. Performing these two translations has in fact been equivalent to a rotation, and not to a translation.

Non-Euclidean translations do not form a group.

This is one of the most characteristic features of non-Euclidean geometry.

The circle and the equidistant may, as we have seen, be described from a common standpoint: a circle is a line which cuts at right angles all straight lines of a pencil of type I, and similarly an equidistant cuts at right angles all straight lines of a pencil of type III.

A line cutting perpendicularly all straight lines of a pencil of type II, that is, all straight lines of a pencil of parallels passing through the same infinitely distant point, will be known as a *horocycle.*

In Fig. 79 we show symbolically a pencil of parallels on the Lobatchevskian plane. Point A traces out a horocycle when travelling perpendicularly to the lines of the pencil. Similarly are generated some lines in physics, e. g. the lines of an electric field cut at right angles the equipotential lines.

Of course, the above intuitive presentation cannot be taken as a proof of the existence of a line which cuts at right angles a pencil of parallels. Moreover, they indicate no concrete method for constructing a point other than A of the horocycle. For both these reasons we should not be satisfied with the above definition of a horocycle but must give it a more constructional character. We shall do this by requiring the straight lines of the pencil to be axes of symmetry of the horocycle and by treating the condition of perpendicular intersection as a theorem to be proved later on.

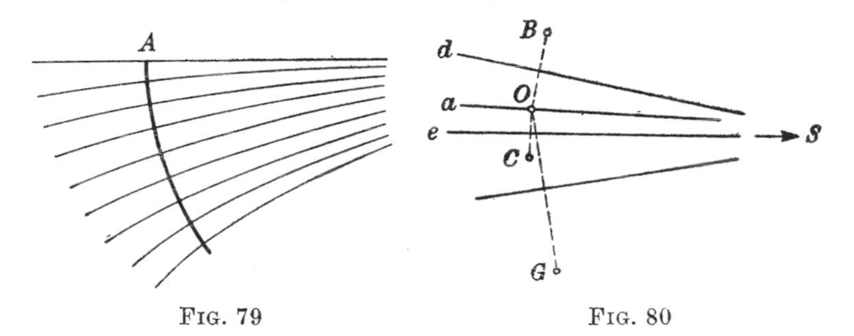

FIG. 79 FIG. 80

If the straight lines $e, d, \ldots$ were axes of symmetry of a horocycle passing through O (Fig. 80), the point symmetrical to O about d would lie on the horocycle as well as the points symmetrical to O about e and an infinite number of lines of the pencil. Thus we define:

A *horocycle* passing through the point O is the line consisting of points symmetrical to O about all straight lines which pass through an infinitely distant point S. Fig. 80 shows how the construction proceeds.

The same construction applied to a pencil of straight lines passing through an ordinary point S obviously gives a circle with centre S. We can therefore refer to a horocycle as a circle with an infinitely distant centre S.

We shall prove that *a straight line passing through the centre of a chord of a horocycle and perpendicular to it will pass through the centre S of our horocycle* (as with a circle), in other words, that it is one of the lines of our pencil (Fig. 81).

This is obvious for chords whose end is O, by the definition of a horocycle; for others, e. g. for BC, it must be proved.

The point B is symmetrical to O about d.

The point C is symmetrical to O about e.

a, b, c, d, e are straight lines of the pencil, that is, they pass through S.

The improper quadrangle $ODSE$ has right angles at D and E. From the last theorem of § 10 (Fig. 58) the diagonal through the vertex O forms with one side of the non-

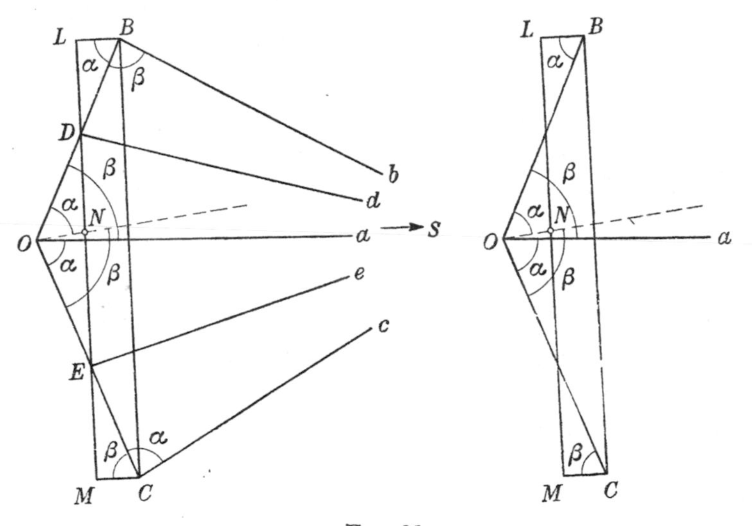

FIG. 81

right angle at O the angle equal to that formed by the perpendicular ON to another diagonal with the second side of our angle. This shows that the angles at O denoted by the letter α are equal, as well as those denoted by β.

The straight line e is the axis of symmetry of the segment OC, the parallel straight lines a and c are symmetrical about it, and therefore the angle marked by a single arc at C is also α.

Similarly the two other angles marked by single arcs at O and B are equal to β.

Let us now apply to the triangle OBC the centres of whose sides are the points D and E the construction of replacing a triangle by a quadrangle of Saccheri (p. 85); that is, let us draw the perpendiculars BL and CM to DE. According to this construction $\sphericalangle DBL = \alpha$, $\sphericalangle ECM = \beta$.

It now appears that we have at the vertices B and C equal angles $\alpha+\beta$. Subtracting from them the equal angles at the vertices B and C of the quadrangle of Saccheri $BLMC$ (Fig. 81) we get two equal angles SBC and SCB.

Therefore, the straight lines b and c form with the segment BC equal angles, which proves that the perpendicular p at the centre of BC (not drawn) is an axis of symmetry of the parallels b and c. Therefore it passes through S and belongs to the pencil (by p. 79). Q. E. D.

It follows from the above theorem that the point O plays no special rôle for the horocycle. In fact, if we were to choose another point on the horocycle, say B, and to construct the point symmetrical to it about the line p of the pencil we should obtain the same point C as before when we were constructing the point symmetrical to O about the line e of the pencil. This means that in both cases we get the same points $C, D, \ldots$ of the horocycle.

The points of a horocycle are equivalent: a point symmetrical to any one of them about an arbitrary line of our pencil will again lie on the horocycle. In other words, *each parallel line of the pencil is an axis of symmetry of the horocycle*—briefly, an axis of the horocycle.

The reader may easily show that the straight line perpendicular at the point A of a horocycle to the axis passing through A cannot be a chord of it, cannot cut it, and therefore it will have only one point in common with it.

Introducing the term "radius of a horocycle" we say that a straight line perpendicular to the radius of a horocycle at the end of that radius is a tangent to the horocycle. In other words, again, a *horocycle cuts perpendicularly all straight lines of a pencil of parallels.*

We have come back to our starting-point but on the way we have proved the existence of a line with this property—we have, namely, demonstrated how to determine any number of its points.

We shall further show that a *horocycle*, like a circle or an equidistant, *may move along itself*. Let h be an arc of a horocycle (Fig. 82) and the half-lines $AS, BS, CS, \ldots$ its axes, i. e. parallel half-lines passing through an infinitely distant point S.

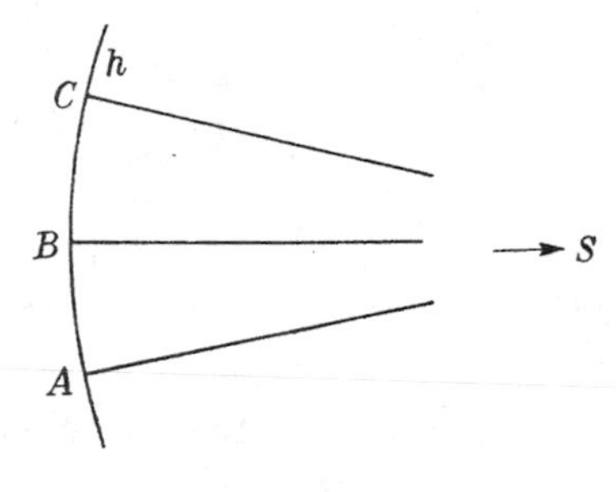

FIG. 82

It is possible to subject the plane to such a motion that any one half-line will be placed upon any other, e. g. AS upon BS. With the plane moving thus parallel lines remain parallel.

Therefore, straight lines parallel to AS will be placed on straight lines parallel to BS, but since BS is parallel to AS lines parallel to AS will remain parallel to it after movement.

In other words, this motion causes merely the interchange of position of straight lines passing through S but their pencil as a whole remains unchanged. It follows that the horocycle will be placed on itself, and in particular the point A on the point B.

We have now seen that motions exist by virtue of which a horocycle slides along itself, just as a circle does when rotated about its centre. We may refer to these motions as rotations of the horocycle about its infinitely distant centre S, when S is an infinitely distant point. These motions form a group.

Point A, when subjected to all the rotations of this group, will trace a horocycle. The analogy with a circle is so far good, but at a certain point will break down.

Let us consider two points A and A_1 (Fig. 83) which trace two concentric horocycles h and h_1 when rotated about S.

Let us translate the whole plane along the straight line a so that A will reach position A_1. The point S, after this translation, will remain unchanged and the pencil of parallel straight lines will be placed upon itself, and for this reason the horocycle h will be placed upon the horocycle h_1. Concentric horocycles are, unlike concentric circles, congruent.

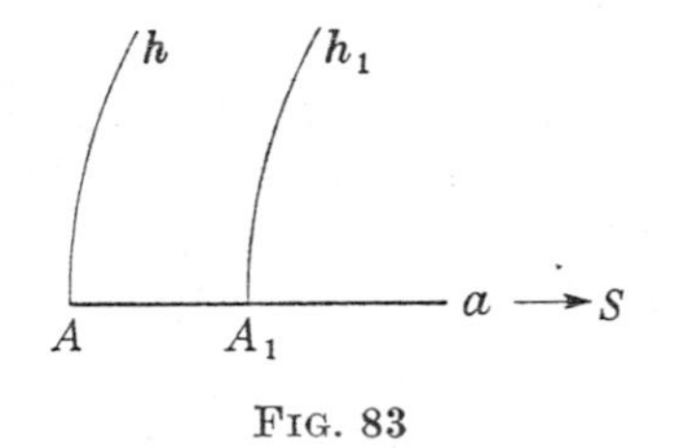

FIG. 83

More generally, since a pencil of parallels with vertex S may be made to coincide with any other pencil of parallels, we get:

All horocycles are congruent.

Therefore, if we consider only its "size and shape", and not its position on the plane, only one horocycle exists, just as only one straight line exists. Thus in Lobatchevskian geometry one might have on one's desk, in addition to the ordinary ruler, a horocyclic ruler, and use it in constructions.

We have achieved a really sensational result. The queen of Euclidean planimetry, the straight line, has a competitor in non-Euclidean geometry, the horocycle, which, like the former, may also slide along itself. It will be no surprise that the study of the properties of this line will take up a central position in our further considerations and permit penetration into the remotest theorems of non-Euclidean geometry.

We shall close this section with a word of praise for the richness and diversity of geometrical forms in Lobatchevskian planimetry. As well as straight lines and circles we find equidistants and horocycles, and new possibilities and fascinating problems appear before us. What an anti-climax that the much poorer geometry of Euclid holds good in this little backwater of the Universe, the Earth!

§ 16. Straight lines and planes in space

In the preceding sections we have presented in brief the changes which are caused by the negation of the axiom of Euclid in the theory of parallels, the theory of the areas of polygons and in the knowledge of circular lines, the latter being closely bound up with certain groups of motions in a plane (the group of translations along a straight line and the groups of rotations about proper and improper centres).

We turn now to space. Numerous theorems of stereometry, for example, those concerned with determining planes through points and straight lines or perpendiculars, mentioned in part on pp. 57-58, do not depend upon the axiom of Euclid and work as well in Lobatchevskian as in Euclidean geometry. Also, the theorem which states that the intersection of a sphere and a plane is a circle (and the similar theorem on two intersecting spheres) is independent of the properties of parallels. These do not exhaust the list of *absolute* theorems of stereometry; this list also includes a number of tests of congruency for trihedral angles, the theorem that the sum of two faces of a trihedron is greater than the third face, and others. We shall have, however, no opportunity to make use of them.

In order to obtain an objective orientation in the mutual positions of straight lines and planes in Lobatchevskian space we shall employ a mapping of the whole of space into a sphere; this proceeds analogously to the mapping of a plane into the interior of a circle (p. 70).

Let us take centre O for the transformation and the acute angle β.

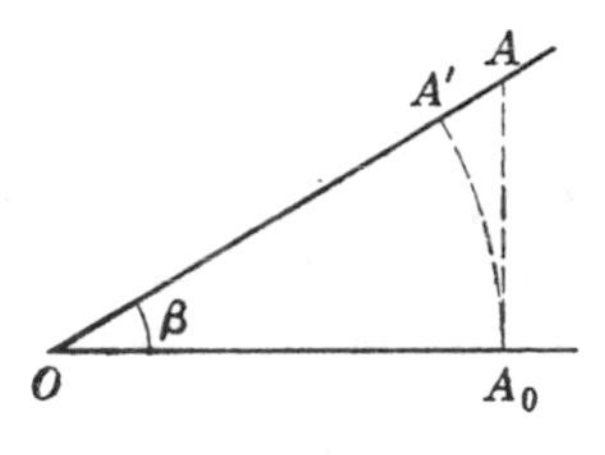

FIG. 84

With each point A of space we associate a point A' lying on the half-line OA at a distance from O equal to the length of the projection of the segment OA onto the straight line which forms the angle β with OA (Fig. 84).

Our planimetric construction repeats in every plane which passes through O: each such plane will be mapped into the interior of a circle with centre O, and each straight line of this plane into a chord of the circle. The radius of this circle will be the same for all planes passing through O, and therefore all these circles will form a sphere with centre O.

Our transformation maps space into the interior of a sphere with centre O; straight lines are mapped into chords, and spheres with centre O into spheres with the same centre.

Intersecting straight lines are mapped into intersecting chords of the sphere, and therefore *every plane* (not only those which pass through O) *will be mapped into the interior of a circle which is a plane section of the sphere.*

In planimetry we called the circle into which the plane had been mapped the map of the plane, so in stereometry we shall call the sphere into which space has been mapped the *model of space.*

We have learnt that in planimetry lines which are parallel (L) in the same direction are, on the map, chords intersecting on the circumference, and so here, on the model, they are chords intersecting on the surface of the sphere; conversely, if straight lines are represented by chords with a common end they will be parallel in the same direction.

Hence, as in planimetry, we obtain the theorem that *if two straight lines are parallel to a third in the same direction, then they will be parallel to each other.*

As we see, the study of parallel lines develops similarly in the stereometry of Lobatchevsky as in planimetry. Similarly, we use the term "infinitely distant point": two parallels a and b are represented in the model by chords cutting at the point S' on the surface of our sphere; a and b are then said to intersect at an infinitely distant point S, and S' to represent on our model that infinitely distant point. On each straight line there are two infinitely distant points.

New, and unexpected, facts appear in the matter of intersecting and non-intersecting planes. Even the definition of parallel planes requires an analysis which is best carried out on the model.

Let us consider two planes α and β which have no points in common. Choosing the centre of transformation on one of them, say β, we get a model of space in which plane β will be represented by a circle β' passing through the centre of the sphere (i. e. its great circle) and plane α will be represented by the circle α' with no interior points in common with the former.

Two cases are possible:

1° The circumferences of the circles α' and β' will have no point in common (Fig. 85).

No chord of circle α' will have any point in common with any chord of circle β', so no straight line in plane

α wil cut or be parallel to any straight line of plane β. The planes α and β will be divergent.

2° The circumferences of the circles α' and β' will have one point P' in common (Fig. 86).

In each of the circles α' and β' there exists a pencil of chords with end P', and for this reason there exist

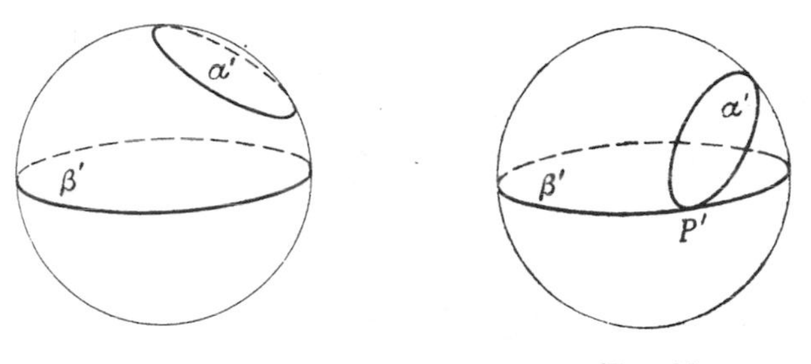

FIG. 85 FIG. 86

on the plane α straight lines which are parallel to certain straight lines on the plane β. These form on α a pencil of parallel lines passing through an infinitely distant point P.

No other straight line lying on α can cut or be parallel to any straight line of β.

Mutatis mutandis the same applies to lines of β.

In this case the planes α and β are said to be *parallel.* In other words, planes are parallel when they have one and only one infinitely distant point in common.

If the planes α and β have two infinitely distant points in common, that is, if the circumferences of the circles α' and β' representing these planes in the model have two common points, then the circles α' and β' will have a common chord (Fig. 87) and the planes α and β a common straight line.

In either of the circles chords passing through P' would represent straight lines parallel to the straight line PQ of intersection of both planes. Similarly with Q'. These and only these straight lines on either plane which are parallel in the same direction to the line of intersection of our planes are parallel to each other.

In conclusion, two planes may, in Lobatchevskian geometry,

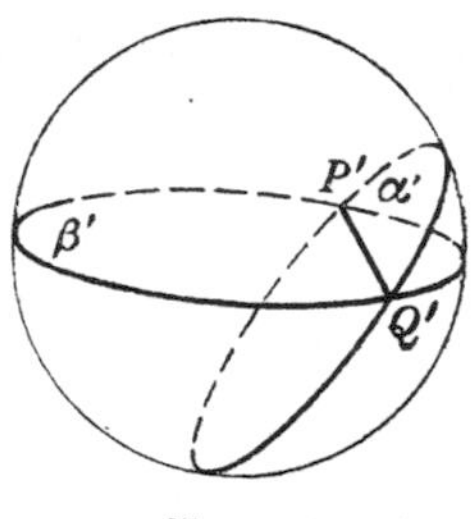

FIG. 87

1° have no infinitely distant points in common (divergent planes),

2° have one such common point (parallel planes),

3° have two such common points (intersecting planes).

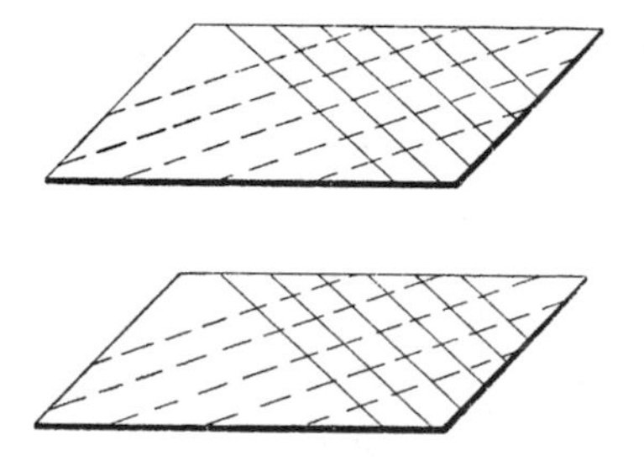

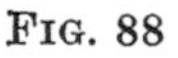

FIG. 88

Comparing these properties with matters in Euclidean geometry, we notice that in the latter

1° divergent planes do not exist at all,

2° on two parallel planes there is an infinite number of pairs of pencils of mutually parallel lines, and not only one (one pencil of a pair lies on the plane α, the other on β and the lines of both are mutually parallel). In Fig. 88 two different pairs of pencils have been distinguished.

Each pencil has one infinitely distant point in common, so therefore two parallel planes may have an infinite number of such points in common.

3° The fewest differences appear with intersecting planes.

The properties of straight lines parallel to a plane may be easily deduced by considering the model. We shall not discuss this topic, but point out that straight lines parallel to a plane and passing through a point C form not a plane but a cone (Fig. 89).

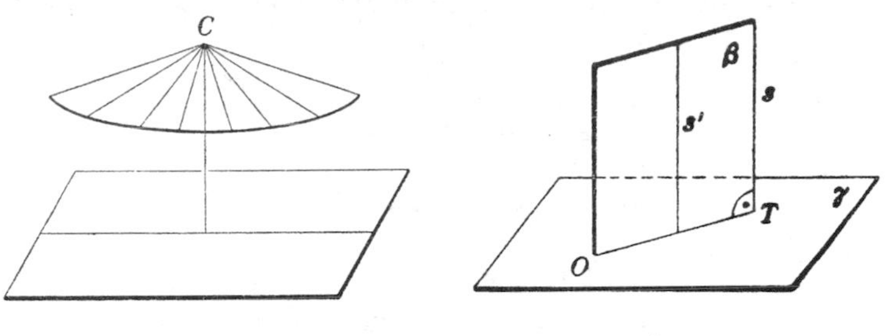

FIG. 89

FIG. 90

In planimetry a right angle one of whose sides passes through the centre of transformation is reproduced on the map as a right angle. This property continues obviously to hold in the stereometric model, so therefore a plane perpendicular to a straight line OA passing through the centre O of transformation will be mapped into a part of a plane perpendicular to the straight line OA'.

Further, one might prove that a straight line s perpendicular to a plane γ which passes through O would be mapped into a chord which would also be perpendicular to γ.

In fact (Fig. 90), the plane γ would be mapped into a circle lying in this plane, line s in plane β into line s' perpendicular (as is line s) to the straight line OT, and therefore also to the plane γ.

Taking advantage of the properties of the model it is possible to prove, and the proof is left to the reader, that:

1. There exists one and only one straight line perpendicular to two divergent planes.

2. The perpendicular projection of one of two divergent planes onto the other is a circle.

3. Two planes parallel to a third may be (a) divergent, (b) intersecting or (c) parallel.

§ 17. The horosphere

The extension of the concept of the equidistant to space presents no difficulties and may be left to the reader. We shall, however, investigate in detail a surface analogous

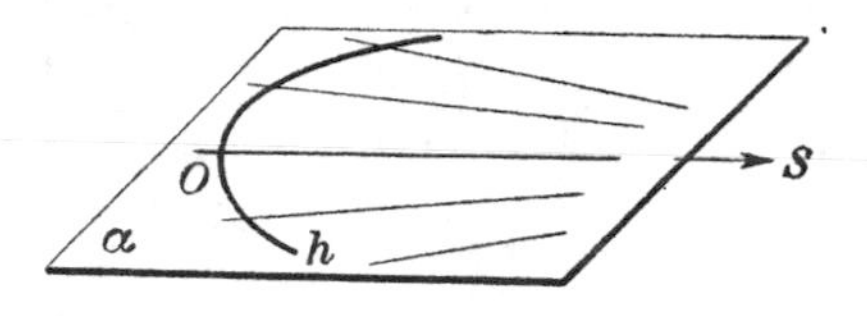

FIG. 91

in the space to the horocycle in the plane and not less interesting. The procedure is similar to that in planimetry with the difference that instead of a plane set of straight lines passing through an infinitely distant point S we now have a set of all straight lines in space which pass through this point, that is, a set of all straight lines parallel in the same direction.

Such a set we call a *bundle* of parallel straight lines.

By *horosphere* we understand a surface consisting of all points symmetrical to a given point O about the lines of a bundle of parallels.

The infinitely distant point S of the bundle will be known as the *centre* of the horosphere and the half-lines connecting the points of the horosphere with S as the *radii* of the horosphere.

Having defined the horosphere let us now examine its properties.

In our bundle we have straight lines parallel to OS and lying in any one of the planes which pass through

OS, let us say, in the plane α (Fig. 91). Points symmetrical to O about these straight lines form the horocycle h.

If the plane α moves round the axis of rotation OS, the horocycle will also move and generate a horosphere. Therefore:

A horosphere is a surface of revolution generated by a horocycle rotating about its axis OS.

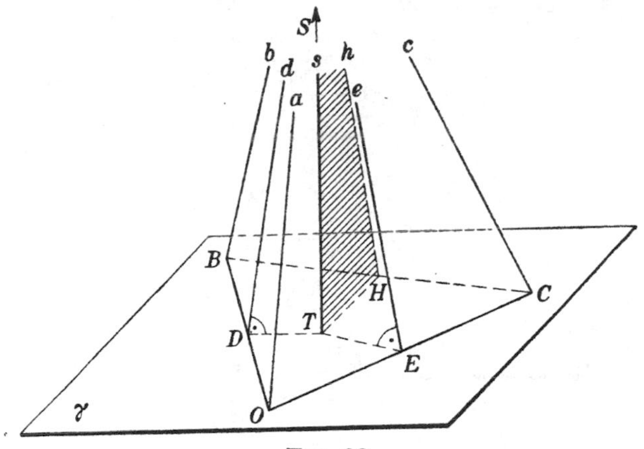

FIG. 92

Let us denote the straight line OS by the letter a and consider two lines d and e of the bundle (Fig. 92). Let the point B be symmetrical to O about d (therefore $OD = DB$) and C about e (therefore $OE = EC$). The points O, B, C, D, E lie in plane γ.

The points O, B, C are points of the horosphere and the segments OB, OC, BC are its chords. The straight lines of the bundle (radii of the horosphere) which pass through the centres of the chords OB and OC are perpendicular to them. We shall show that the same applies to the chord BC, or that

The radius of a horosphere which passes through the centre of a chord of it is perpendicular to the chord.

PROOF. Let us draw the radii of the horosphere through B and C. Turning from Fig. 92 to its image in the model, when O has been chosen as the centre of transformation, we find:

a, d, e are mapped into the chords passing through a certain point S' on the surface of the sphere (Fig. 93), and the right angles at D and E are mapped into right angles.

Let us draw from S' the straight line $S'T'$ (s') perpendicular to the plane γ.

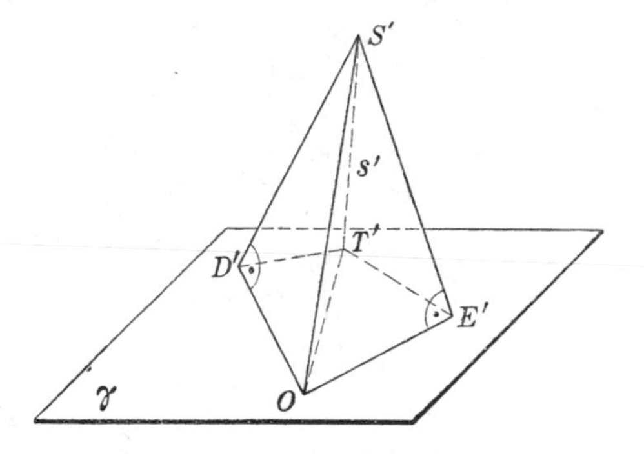

FIG. 93

By the theorem on three perpendiculars (p. 57) we get

$$T'D' \perp OD' \quad \text{and} \quad T'E' \perp OE'.$$

In other words, the straight lines perpendicular to OE' and OD' at their ends cut at the point T'.

Therefore, and because of the properties of the model which are given at the end of § 16, it follows that the line s (ST), which is represented by s' (Fig. 92), is perpendicular to γ and that the axes of symmetry of the sides OB and OC of the triangle OBC (i. e. the straight lines perpendicular to these sides at their centres) intersect at the point T equidistant from the points O, B, C; consequently T lies on the axis TH of the third side BC (H is the centre of BC).

Draw radius h of the horosphere through the point H. The lines s and h, being parallel, lie in one plane (shaded in Fig. 92).

This plane is perpendicular to the plane γ, since it passes through s and is perpendicular to BH because the latter segment is perpendicular to the edge TH of both planes.

The segment BC is perpendicular to the plane on which h lies, so BC is perpendicular to h.

We have proved that the radius of the horosphere passing through the centre of the chord BC is perpendicular to this chord. Q. E. D.

Next we shall show, exactly as on p. 107, that the point O plays no special rôle in the horosphere. Choosing instead the point B and finding the points symmetrical to B about the lines of the bundle we should obtain the same horosphere.

All points of the horosphere are equivalent. Each line of the bundle is therefore an axis of revolution of the horosphere, so that we are entitled to consider the horosphere as a sphere with an infinitely distant centre.

We see, without difficulty, that a plane perpendicular to the radius of a horosphere at the end of this radius is tangent to the horosphere. Therefore, a horosphere is a surface which cuts at right angles all lines of a bundle.

Further properties of the horosphere follow directly from the corresponding properties of the horocycle and from the fact that the horosphere is a surface of revolution generated by a horocycle moving about its axis:

A horosphere is determined by its centre and one of its points.

Every plane which passes through the radius of a horosphere is a plane of symmetry of the horosphere.

All horospheres are congruent.

Thus, if we ignore position, only one horosphere exists, just as only one plane exists.

Let us now examine the structure of the horosphere. It is an infinite surface and an infinite number of horocycles lie on it.

Each plane passing through a radius of a horosphere cuts it in a horocycle. In Fig. 94 the radius has been denoted by a, the plane by α.

The radii passing respectively through the points A and B of a horosphere are parallel and determine a plane α which intersects the horosphere along a horocycle connecting A and B.

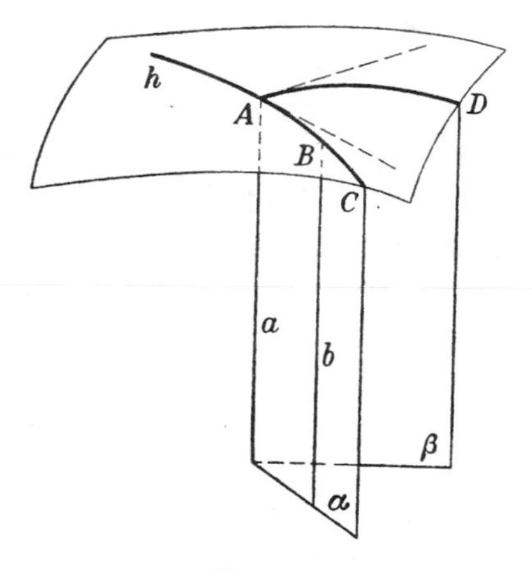

FIG. 94

Two points on a horosphere may be joined by one and only one horocycle which lies on this horosphere.

A certain arc on this horocycle may be chosen as a unit and a certain point, say A, may be chosen as the initial one. Then we can assign to every other point C of the horocycle a real number $x =$ the measure of the arc AC in terms of the chosen unit. The points of the horocycle are arranged just as the points of a straight line.

A horocycle divides a horosphere into two parts.

Two "half-horocycles" originating at point A (AC and AD in Fig. 94) form an angle. This angle, according to the general convention, is to be taken as the angle between the straight lines tangent at A to both horocycles.

These tangents are perpendicular to the common axis a of both horocycles and lie in the planes α and β of the horocycles, so thus the angle between them is the plane angle of the dihedral angle formed by these two planes. It is therefore equal to the angle through which one plane should be rotated so as to coincide with the other.

The angle between two horocycles is equal to the angle of rotation which brings one of them to coincide with the other. During this rotation the horosphere would be slid-

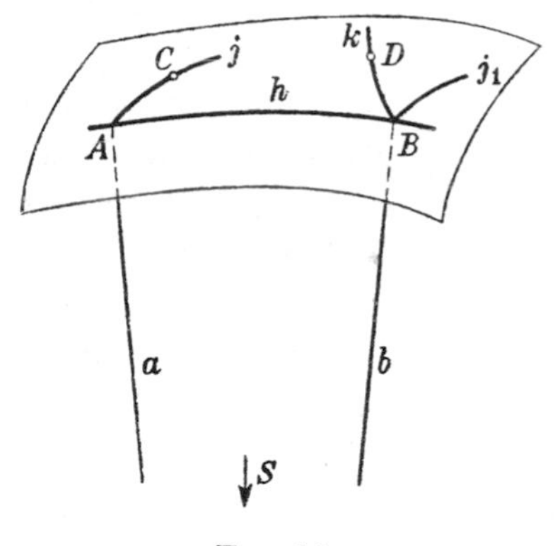

FIG. 95

ing over itself. So we say that *a horosphere may be rotated about its point A.*

A horosphere may also be translated along itself so that a certain horocycle would slide along itself and a point A on it be placed upon any other point B, also on it.

Let us in fact move the horocycle h on its plane (Fig. 95) round its centre S and let us suppose that a horosphere is rigidly attached to it. If point A takes up the position B, the horosphere which passes through A and has centre S will coincide with the horosphere which passes through B and has the same centre. Therefore we have:

Among the motions of a horosphere over itself there are rotations about any of its points and translations along any horocycle.

Therefore any half-horocycle j (Fig. 95) may be made to coincide with any other one k. In order to do this it would suffice first to translate the horosphere along the horocycle h so as to place the point A on B and the horocycle j on the horocycle j_1, and then to rotate the horosphere through a suitable angle about B. Thus the arc AC would take up the position of the equal arc BD.

Finally, we may transform the horosphere by symmetry with respect to the horocycle h, since the plane of this horocycle is a plane of symmetry of the horosphere. Symmetrical figures have equal sides and equal angles.

All the properties of the horosphere so far given are the same as those of the plane, with the only difference that instead of "straight line" we have "horocycle", and instead of "segment" we have "arc" of a horocycle. On the horosphere one may examine horocyclic triangles, polygons and circles and also other figures analogous to those of planimetry.

All the sides and angles of a triangle ABC are equal to the corresponding sides and angles of the triangle $A_1B_1C_1$ if $\measuredangle A = \measuredangle A_1$, $AB = A_1B_1$, $AC = A_1C_1$.

In fact, as we have seen, a horosphere may be moved so as to make the "segment" AB coincide with A_1B_1. When this happens the angle A either coincides with angle A_1 or with angle A_2 symmetrical to it. In the latter case symmetry with respect to A_1B_1 would transform the angle A_2 into A_1. Because of $AC = A_1C_1$ the vertex C will eventually be placed upon C_1, and therefore we have $\measuredangle B = \measuredangle B_1$, $\measuredangle C = \measuredangle C_1$, $BC = B_1C_1$.

Hence the tests of congruency for horospheric triangles are exactly the same as for plane triangles.

It follows from the above remarks that all the fundamental properties (axioms) of the horosphere which refer to position, arrangement and congruence are precisely the same as the corresponding axioms of planimetry, providing that the term "straight line" is replaced by "horocycle" and "segment" by "arc of a horocycle". Consequently the same also applies to the arguments resting on these axioms. On the horosphere all the arguments and theorems of absolute geometry hold true.

We do not yet know how the matter of intersecting or non-intersecting horocycles will look on the horosphere. Let us return to Fig. 92, in which the points O, B, C lie on the horosphere and the line s is perpendicular to the plane of the triangle OBC. The straight line s is one of a bundle of radii of the horosphere and its point T is equidistant from the points O, B, C. Therefore, rotating the entire figure about s, we may make TO coincide with TB. When this happens the horosphere will slide over itself and the point O will trace a circle on the horosphere. This circle passes through O, B and also C. Therefore: through any three points of a horosphere which do not lie on one horocycle there passes a circle. But we have seen (p. 90) that if the axiom of Euclid does not hold not any triangle may be inscribed in a circle. Thus this axiom holds on the horosphere together with all its consequences: on a horosphere the sum of the angles of a triangle is 180°; the opposite sides of a parallelogram are equal, the theorems of Thales, of Pythagoras are true, and so forth.

The surface of the horosphere is governed by Euclidean geometry.

This theorem, which was discovered by Lobatchevsky and Bolyai, is one of the most interesting and most important of the non-Euclidean geometry. The whole of chapter III of this book will be based on it.

The following paradox appears: if Eulidean planimetry is not true, it will be true nonetheless. If, indeed, it is not true on the plane, it will hold good on the horosphere.

This fact preserves intact the ordinary geometrical theory of trigonometrical functions: we may define sine, cosine and tangent as the ratios of the sides of right-angled triangles on the horosphere and derive the properties of these functions exactly as in elementary trigonometry. On p. 29 we were complaining that together with the axiom of Euclid plane trigonometry would also fail; our grievance vanishes now, for we have horospheric trigonometry, and the study of the properties of the actual functions is unaltered.

A question about the converse situation naturally comes up: does there exist in Euclidean space—if we assume that Euclidean geometry is valid on the plane—a surface and such a set of lines on it that theorems on these lines will be the same as those of non-Euclidean geometry, provided that the words "straight lines" are replaced by these lines?

The answer is "yes", though not without limitation. A surface like this, called a *pseudosphere*, was investigated by Beltrami about 1870. It was an essential contribution to the struggle against the old opinions which considered the negation of the axiom of Euclid to be pure nonsense. If the theorems of non-Euclidean geometry do, in fact, accurately describe the conditions in force on some surface, there is nothing stupid inherent in them, nothing nonsensical.

Unfortunately, an examination of the pseudosphere would require finer mathematical methods, and we must pass it by in the present book.

CHAPTER III

FURTHER DEVELOPMENT OF THE THEORY

§ 18. Arcs of concentric horocycles

One may define the length of an arc of a horocycle in the same way as the length of the arc of a circle—that is, as the limit of the lengths of polygonal lines which have been inscribed in the arc, or as the upper limit of their lengths. We shall not go here into details but merely take it for granted that every arc of a horocycle has a definite "length" and that if a number of arcs be put together their lengths are to be added.

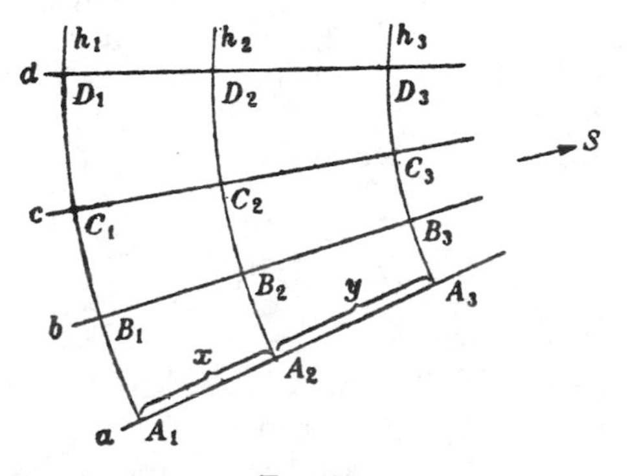

Fig. 96

In Fig. 96 are shown coplanar horocycles h_1, h_2, h_3 with centre S, from which the parallel lines passing through S cut off arcs A_1B_1, B_1C_1 etc.

If $\smile A_1B_1 = \smile B_1C_1$ a certain rotation about S would make the former arc coincide with the latter, therefore also the line a with b and b with c. Consequently the arc A_2B_2 would move over to the position B_2C_2, whence $\smile A_2B_2 = \smile B_2C_2$.

If the arcs on the horocycle h_1 are equal the corresponding arcs of the concentric horocycle h_2 will also be equal.

It follows that if the ratio of the arcs A_1C_1 and C_1D_1 is a rational number $\frac{m}{n}$, i.e. if A_1C_1 consists of m equal parts and C_1D_1 of n parts of the same size, then matters are not altered with the arcs A_2C_2 and C_2D_2 of the horocycle h_2. Therefore $\frac{\smile A_2C_2}{\smile C_2D_2}$ is also $\frac{m}{n}$. Hence

$$\frac{\smile A_1C_1}{\smile C_1D_1} = \frac{\smile A_2C_2}{\smile C_2D_2}.$$

Similarly as with the proof of the theorem of Thales in Euclidean geometry we can prove the above formula also in the case where $\frac{\smile A_1C_1}{\smile C_1D_1}$ is irrational.

Therefore, *the corresponding arcs of concentric horocycles are proportional.*

Transposing, we obtain

$$\frac{\smile A_1C_1}{\smile A_2C_2} = \frac{\smile C_1D_1}{\smile C_2D_2} = \text{const.}$$

This constant, which is greater than one, since $A_2C_2 < A_1C_1$, depends upon which horocycles h_1 and h_2 are taken. A figure consisting of two concentric horocycles is characterised by the distance between them, that is, by the segment $A_1A_2 = x$. Therefore the constant value of the quotient $\frac{\smile A_1C_1}{\smile A_2C_2}$ is a certain function of this variable:

$$\frac{\smile A_1C_1}{\smile A_2C_2} = f(x).$$

$f(x)$ is an increasing function, since on moving point A_2 to the right of A_1 the parallels a and c, of the bundle would approach and A_2C_2 would diminish.

Let $A_2A_3 = y$. According to the above statement the ratio of the corresponding arcs A_2C_2 and A_3C_3 of the horocycles distant by y is $f(y)$, i. e.

$$\frac{\smile A_2C_2}{\smile A_3C_3} = f(y).$$

Multiplying these two ratios we get

$$\frac{\smile A_1C_1}{\smile A_3C_3} = f(x)f(y).$$

But since the horocycles h_1 and h_2 are distant by $x+y$, the ratio of their corresponding arcs is

$$\frac{\smile A_1C_1}{\smile A_3C_3} = f(x+y).$$

Thus we have for every positive x and every positive y

$$f(x+y) = f(x)f(y).$$

It is easy to see that the formula remains valid also for $x = 0$ and for $y = 0$. The function $f(x)$ must satisfy it. We now face the problem of *determining the increasing positive function of one variable* $f(x)$ *satisfying the functional equation*

$$f(x+y) = f(x)f(y)$$

for all non-negative values of x *and* y.

It appears that the solution of our problem depends on the value taken by the function at $x = 1$. When this value is given, when, for example, it is equal to a $(a > 0)$, the function will be uniquely determined.

Let

$$f(1) = a,$$

then

$$f(2) = f(1+1) = f(1)f(1) = a.a = a^2.$$

Similarly

$$f(3) = a^3, \ldots, f(k) = a^k \quad (k \text{ a positive integer}).$$

Let m be a positive integer. We have

$$f(1) = f\underbrace{\left(\frac{1}{m} + \frac{1}{m} + \ldots + \frac{1}{m}\right)}_{m \text{ terms}} = \underbrace{f\left(\frac{1}{m}\right) f\left(\frac{1}{m}\right) \ldots f\left(\frac{1}{m}\right)}_{m \text{ factors}},$$

owing to the functional equation. Therefore

$$\left[f\left(\frac{1}{m}\right)\right]^m = a, \quad \text{i. e.} \quad f\left(\frac{1}{m}\right) = a^{\frac{1}{m}}.$$

We have found that the function $f(x)$ assumes the same values at positive rational x as the function a^x; the number a must exceed 1, since in the contrary case $f(x)$ would not increase.

Also with irrational x the function $f(x)$ will assume the value a^x, since otherwise it would obviously not increase.

Hence we have for all positive x

$$f(x) = a^x.$$

This function satisfies the conditions formulated above.

We have solved our problem and found that the function in question is a^x, i. e. the exponential function with base a.

We generally use in mathematics an exponential function whose base is a certain irrational number (whose definition we cannot go into here); its approximate value is

$$2{\cdot}7182818\ldots$$

This number will be denoted by e and the exponential function by e^x.

The constant a appearing in the formula $f(x) = a^x$ may be put in the form $a = e^b$, where $b = \log_e a$. b will be in further developments a very small number, so we shall express it $b = \dfrac{1}{k}$.

In that case

$$a = e^{\frac{1}{k}}, \quad f(x) = e^{\frac{x}{k}}.$$

Let us return to the arcs of the concentric horocycles. We shall express the formula $\frac{A_1C_1}{A_2C_2} = f(x)$ as follows:

The ratio of the corresponding arcs of two concentric horocycles distant by x is equal to $e^{\frac{x}{k}}$, k being a certain positive constant.

The value of this constant depends upon the choice of the unit of measurement. If we were to diminish this unit twice, the ratio of the arcs would not alter, whence neither would the function $e^{\frac{x}{k}}$, but as x would increase twice, the value k would also do so. If $x = k$, then the ratio of the arcs of concentric horocycles will be equal to e, so therefore k is the distance between two concentric horocycles, when the ratio of their corresponding arcs is e. The value k expresses the l e n g t h of a certain definite segment.

Let us consider a number of concentric horocycles at distance x from each other. Let the corresponding arcs of them have lengths $a_1, a_2, a_3, \ldots$ The ratio of each one to the next is $e^{\frac{x}{k}}$, whence $a_1, a_2, a_3, \ldots$ form a geometrical progression with quotient $\frac{1}{e^{\frac{x}{k}}} = e^{-\frac{x}{k}}$.

If the first arc is 1, the second 0·9, then the third one will be $(0{\cdot}9)^2$, the fourth $(0{\cdot}9)^3$ etc.

§ 19. Hyperbolic functions

The theorem given in the last section explains why we so frequently find in the formulae of non-Euclidean geometry the exponential function e^y. Some expressions depending upon it are of great importance and have their own names.

The function $\frac{1}{2}\left(e^y + \frac{1}{e^y}\right) = \frac{1}{2}(e^y + e^{-y})$ is called the *hyperbolic cosine* of y and is denoted by $\cosh y$.

The function $\frac{1}{2}\left(e^y - \dfrac{1}{e^y}\right) = \frac{1}{2}(e^y - e^{-y})$ is called the *hyperbolic sine*, in short $\sinh y$ [1].

The following relationship exists between these functions:

$$\cosh^2 y - \sinh^2 y = 1, \qquad (1)$$

which may be verified by substituting the given values of the hyperbolic cosine and hyperbolic sine. This relationship resembles the equality

$$\cos^2 y + \sin^2 y = 1.$$

Using it we may, as in trigonometry, evaluate one of the functions $\cosh y$ or $\sinh y$, given the other.

The introduction suggests itself of the hyperbolic tangent by the formula

$$\tanh y = \frac{\sinh y}{\cosh y}.$$

Again, as in trigonometry, $\cosh y$ and $\sinh y$ may be determined, when given $\tanh y$.

There are further analogies between the hyperbolic and trigonometrical functions; the most important of them are those involved with the formulae for $\cosh(y+z)$ and $\sinh(y+z)$. The first of them may be obtained on considering the expression

$$\cosh y \,.\, \cosh z + \sinh y \,.\, \sinh z$$
$$= \frac{e^y + e^{-y}}{2} \cdot \frac{e^z + e^{-z}}{2} + \frac{e^y - e^{-y}}{2} \cdot \frac{e^z - e^{-z}}{2}.$$

An easy calculation shows that the right-hand side of the above equation is $\dfrac{e^{y+z} + e^{-y-z}}{2}$ which, according to the definition of the hyperbolic cosine, is precisely

(1) The origin of the term "hyperbolic" is explained in the Appendices.

$\cosh(y+z)$. Therefore:

$$\cosh(y+z) = \cosh y \,.\, \cosh z + \sinh y \,.\, \sinh z. \qquad (2)$$

In the analogous trigonometrical formula the sign $-$ appears in the place of $+$.

It is not difficult to discover the formula for $\sinh(y+z)$, which has the same form as the corresponding trigonometrical one.

We shall not examine in detail the properties of the hyperbolic functions since we do not need them here; we shall only discuss briefly the values assumed by these functions for positive values of y.

1° If $y = 0$, $\sinh y = 0$.

As y increases, then in the formula $\sinh y = \dfrac{e^y - e^{-y}}{2}$ the minuend increases, the subtrahend decreases, and so $\sinh y$ increases. As the function e^y increases very rapidly, the same property will be possessed by $\sinh y$.

2° If $y = 0$, then $\cosh y = \dfrac{1+1}{2} = 1$.

It follows from the relationship $\cosh^2 y = 1 + \sinh^2 y$ that $\cosh y$ will also increase as y does, and assume very large values.

3° If $y = 0$, then $\tanh y = 0$.

It follows from the formula

$$\tanh y = \frac{e^y - e^{-y}}{e^y + e^{-y}} = \frac{1 - e^{-2y}}{1 + e^{-2y}}$$

(we have divided the numerator and the denominator by e^y) that when y increases the subtrahend in the numerator will decrease, so the numerator will increase and the second term in the denominator will decrease, so the denominator will decrease, with the final result that the quotient will increase. For a sufficiently large y the value of e^{-2y} is very small and the value of $\tanh y$ is near to unity; we shall write this down: $\tanh y \to 1$ as $y \to \infty$.

The introduction of hyperbolic functions is useful but not essential; we could discard the symbols $\sinh y$ and $\cosh y$ and write instead $\frac{e^y - e^{-y}}{2}$ and $\frac{e^y + e^{-y}}{2}$. They are, however, convenient, and the analogies with the trigonometrical functions facilitate working with them. The approximate values of the hyperbolic functions are given in special tables, of which we have included an except at the end of the present book. We may point out that, generally speaking, a table of decimal logarithms would suffice. This may be shown by two examples; we shall not use them, though, later.

EXAMPLE 1. Find $\cosh y$, given $y = 1{\cdot}8$.

Let us write $z_1 = e^y$ and $z_2 = e^{-y}$.

$$\log z_1 = 1{\cdot}8 \log e = 1{\cdot}8 \,.\, 0{\cdot}4343 = 0{\cdot}7817, \qquad z_1 = 6{\cdot}049;$$

$$\log z_2 = -0{\cdot}7817 = \bar{1}{\cdot}2183, \qquad z_2 = 0{\cdot}1653,$$

$$\cosh 1{\cdot}8 = \frac{6{\cdot}049 + 0{\cdot}1653}{2} = 3{\cdot}107.$$

EXAMPLE 2. Find y, given $\sinh y = 2$.

By definition $\sinh y = \frac{e^y - e^{-y}}{2}$. Let us write $e^y = z$.

Hence

$$\frac{z - \frac{1}{z}}{2} = 2 \quad \text{and} \quad z^2 - 4z - 1 = 0.$$

Since $z = e^y > 0$, we get

$$z = e^y = 2 + \sqrt{5} = 4{\cdot}2361.$$

Taking logarithms

$$y \log e = \log 4{\cdot}2361, \qquad y = 0{\cdot}6270 \,.\, 2{\cdot}3026\,(^1) = 1{\cdot}444.$$

---

(1) The value $\frac{1}{\log e}$ is given in the tables.

Until now our remarks have not exceeded the scope of elementary mathematics. However, in the following sections we shall be in need of some theorems on the quantity e which cannot be proved here as we have not defined this number precisely.

The reason why the number e is so particularly convenient as the base of the exponential function is that for very small values of y we have "approximately"

$$e^y = 1+y,$$

or more exactly that the quotient $\frac{e^y-1}{y}$ tends to 1 as y tends to zero. Other functions, for instance, 2^y, do not possess this property. 2^y, when y is small, assumes approximately the value $1+ay$, where a is a certain irrational number less than 1.

A better approximation to e^y than $1+y$ is given by $1+y+\frac{1}{2}y^2$; e. g., when $y = 0{\cdot}1$, then $e^y = 1{\cdot}1052$, $1+y = 1{\cdot}1$, but $1+y+\frac{1}{2}y^2 = 1{\cdot}105$. It is possible to find still closer approximations to the function e^y as also to $\sinh y$ and $\cosh y$.

Later on we shall make use of the theorem which follows from them, that *for sufficiently small values of y (for $y < 1$) the following formulae are approximately true*:

$$\sinh y = y+\tfrac{1}{6}y^3, \tag{3}$$

$$\cosh y = 1+\tfrac{1}{2}y^2. \tag{4}$$

In both cases the given values are too small and the error is less than $\frac{1}{20}y^4$, e. g. $\cosh\frac{1}{2} = 1+\frac{1}{8}$ with an error of less than $\frac{1}{320}$.

It follows from the above theorem that

$$\frac{\sinh y}{y} = 1+\frac{1}{6}y^2+\ldots$$

tends to 1 as y tends to 0.

§ 20. The function of Lobatchevsky

In this section we shall settle the fundamental question of the nature of the Lobatchevskian function Π. We recall that the angle of parallelism α (Fig. 97) is a function of the distance q ($k \| j$, $BA \perp j$):

$$\alpha = \Pi(q).$$

The problem of discovering the relationship between the angle α and the segment q may be formulated as a problem about horocycles.

Let us draw a horocycle with centre S through the point B. The segment AB is one-half of the chord of our horocycle, and the angle α is the angle between the chord and the radius of the horocycle.

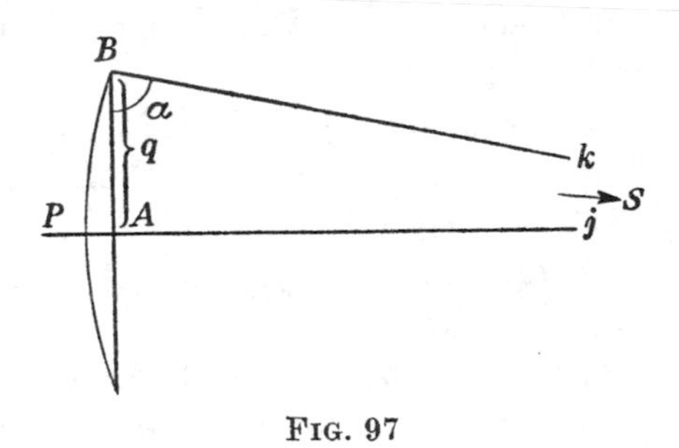

FIG. 97

Now our problem consists on the determination of the relationship between the chord of the horocycle and the angle formed by this chord with the radius.

In the simplest case $\alpha = 45°$, the tangent BD to the horocycle will be symmetrical about AB to the straight line k, and therefore it will also be parallel to the line j. Let us denote by m the length of the arc of the horocycle between the vertex of the right angle DBk down to the straight line j which is parallel to both sides of the right angle (Fig. 98).

The basic idea of the argument which we shall now give originates from Lobatchevsky and rests on two foundations:

1° On the theorem that Euclidean geometry holds good on the horosphere, and

2° on the theorem on the arcs of concentric horocycles, and so on the properties of horocycles.

Let us consider the triangle ABC with the right angle at B. We imagine the plane I of our triangle as being

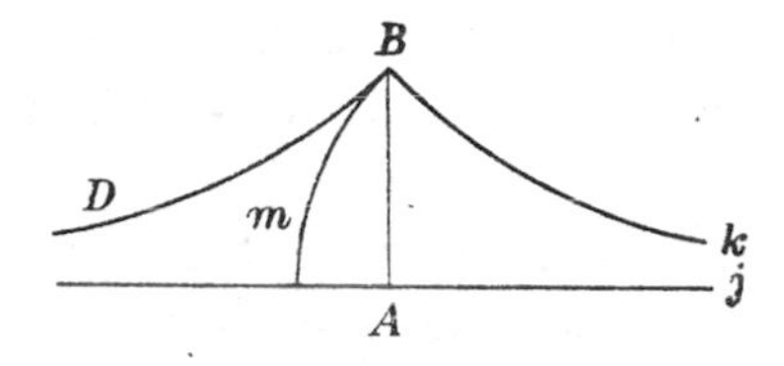

FIG. 98

horizontal (Fig. 99). The vertex C may be either a proper or an infinitely distant point. The reasoning is only slightly different in the two cases and we shall discuss here the second possibility.

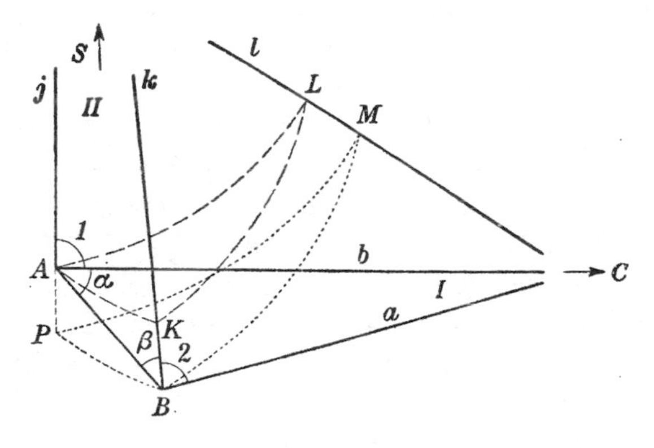

FIG. 99

Let us draw through A a straight line j perpendicular to the plane I of the triangle ABC, and through the points B and C the lines k and l parallel to j (k and l will cut j at the infinitely distant point S). We shall denote by α

the angle CAB. Before we start the actual proof let us take a look at the configuration of Fig. 99.

The line l is parallel to both sides of the right angle *1*. The plane II of the lines j and k is perpendicular to the plane I and intersects it along AB. The straight line BC is perpendicular to AB, and therefore perpendicular to the plane II. Consequently the angle *2* is right and the line l is parallel to both its sides. Finally, since a is perpendicular to the plane II, so also will be the plane of lines k and a.

We have seen that several angles in Fig. 99 are right; we may now add that the angles denoted by α and β are equal. β is, in fact, the angle of parallelism corresponding to the distance AB, since j and k are parallel. On the other hand b is parallel to a, so that α is the angle of parallelism corresponding to the same distance AB; therefore

$$\alpha = \beta .$$

Having now described the figure let us now turn to the essential part of the argument. Let us draw two horospheres with centre S, one through A, the other through B. The former will cut the straight lines j, k, l at the points A, K, L, the latter at the points P, B, M.

On the first horosphere we have the horospheric triangle AKL, and on the other the triangle PBM. First, let us consider triangle AKL. The angle K is right, since, as we have pointed out, the plane of the lines a and k is perpendicular to the plane II. Similarly we show that the angle A is equal to α.

Since Euclidean geometry governs the horosphere, we get

$$\smile KL = \smile AL \sin\alpha , \qquad \smile AK = \smile AL \cos\alpha . \tag{5}$$

Now let us consider the triangle PBM. Its horosphere is distant from the horosphere of triangle AKL by the segment $AP = BK$. We apply the theorem of p. 129 to the arcs KL and BM of the concentric horocycles:

$$\frac{\smile BM}{\smile KL} = e^{\frac{BK}{k}},$$

and substitute for $\smile KL$ the value obtained previously; then

$$\frac{\smile BM}{\smile AL \sin \alpha} = e^{\frac{BK}{k}}.$$

Let us note, that angles *1* and *2* are right. In both angles the arc of the horocycle (AL and BM respectively) runs

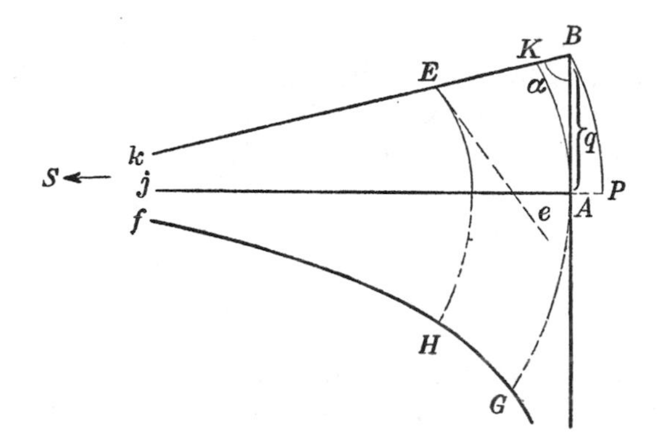

Fig. 100

from the vertex to the straight line which is parallel to both sides of the angle. Therefore both these arcs are equal to m, if we use the notation of Fig. 98, and the above formula takes the form

$$\frac{1}{\sin \alpha} = e^{\frac{BK}{k}}.$$

For clarity's sake let us draw separately the figure (Fig. 100) which has appeared in the plane *II*, and which differs in no way from Fig. 97.

Let us take $BE = BA$ and construct the straight line e perpendicular to k. Then we have

$$BA \| e,$$

since α is the angle of parallelism corresponding to the segment $BA = BE$.

We now produce BA and construct the line f parallel to both j and BA in the direction indicated in the figure. Then we have also

$$f \| k, \quad \text{since} \quad k \| j,$$

$$f \| e, \quad \text{since} \quad e \| BA.$$

Let us produce the horocycle KA to the point G of intersection with f; let us, further, draw through E the horocycle EH with the same centre S. This is done in order to get two concentric horocycles and make use of the theorem on their arcs:

$$e^{\frac{EK}{k}} = \frac{\smile KG}{\smile EH} = \frac{\smile AK + \smile AG}{\smile EH}.$$

Note that $\smile AG$ runs from the vertex A of the right angle down to the line f which is parallel to both sides of it, so $\smile AG = m$ (notation of Fig. 98). On the other hand EH begins also at the vertex of a right angle, viz. E, and continues down to f, which is parallel to both sides of the angle. Thus $\smile EH = m$.

Substituting this into the preceding formula, and recalling (5), we obtain

$$e^{\frac{EK}{k}} = \frac{\smile AK + m}{m} = \frac{m \cos \alpha + m}{m},$$

$$1 + \cos \alpha = e^{\frac{EK}{k}}.$$

Multiplying this formula by the previous one $\left(\frac{1}{\sin \alpha} = e^{\frac{BK}{k}}\right)$

we get:

$$\frac{1+\cos\alpha}{\sin\alpha} = e^{\frac{BA}{k}}.$$

But, since in the previous notation (Fig. 97) $BA = q$, we have

$$\frac{1+\cos\alpha}{\sin\alpha} = e^{\frac{q}{k}}.$$

Using the identities

$$1+\cos\alpha = 2\cos^2\tfrac{1}{2}\alpha,$$

$$\sin\alpha = 2\sin\tfrac{1}{2}\alpha\cos\tfrac{1}{2}\alpha$$

we come to the final result

$$\cot\tfrac{1}{2}\alpha = e^{\frac{q}{k}}. \tag{6}$$

The relationship between the segment q and the angle of parallelism α has been found. Formula (6) solves the fundamental problem of Lobatchevskian geometry and determines the nature of the function Π. Namely:

The angle of parallelism corresponding to the distance q is twice as great as the acute angle whose cotangent would be $e^{\frac{q}{k}}$.

Using the above formula we may determine $\frac{q}{k}$ when given the angle α and conversely we may determine α when given the ratio $\frac{q}{k}$.

1° Let α equal, say, 89°. Using the tables of decimal logarithms we obtain

$$\frac{q}{k} = \frac{1}{\log e}\log\cot 44{\cdot}5° = 2{\cdot}3026\log\cot 44{\cdot}5° = 0{\cdot}0175.$$

Therefore, the angle of parallelism is 89° when the distance q is 0·0175 k. When $\alpha = 89°59'58''$ this distance is 0·00001 k.

2° Fig. 101 represents a graph of α as a function of the variable $r = \frac{q}{k}$; its points have been determined by formula (6).

We have said on p. 129 that the constant k of our formula is the length of a certain definite segment, namely, the distance between two concentric horocycles whose

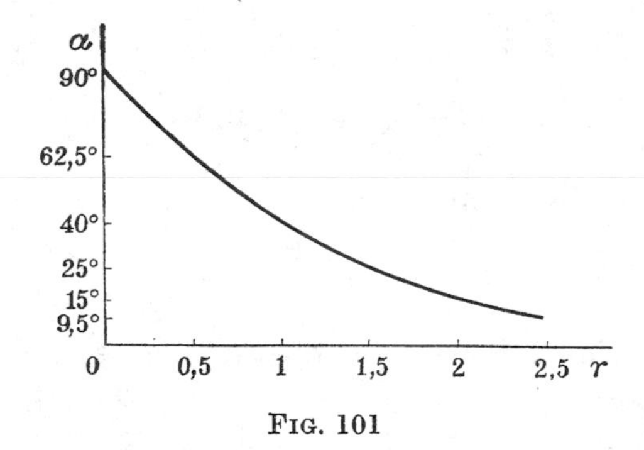

FIG. 101

corresponding arcs have the ratio equal to e:1. We may now describe this segment in yet another way, namely, for $q = k$,

$$\cot \tfrac{1}{2}\alpha = e = 2{\cdot}71828\ldots,$$

whence

$$\alpha = 40°23'42'',$$

and k is the segment corresponding to the angle of parallelism of the above magnitude. Therefore this segment may be determined in a basically simple manner: it is necessary to go so far away from a given straight line that the angle of parallelism will be $40°23'$... Naturally, an experiment like this is not practically possible in the space which surrounds us. We shall return later to this question.

The expression $\frac{q}{k}$ appearing in the formula $\cot\frac{1}{2}a = e^{\frac{q}{k}}$ is the ratio of the segment q to the segment k. In the formulae of non-Euclidean geometry we always meet with ratios of segments to the segment k. These ratios might be called *measures* of these segments with k as the unit.

We have made $\cot\frac{1}{2}a$ depend upon q. It is not difficult to find the values of the functions $\sin a$ and $\cos a$; it suffices to make use of trigonometrical formulae for a double angle:

$$\cos a = \cos^2\tfrac{1}{2}a - \sin^2\tfrac{1}{2}a = \frac{\cos^2\frac{1}{2}a - \sin^2\frac{1}{2}a}{\cos^2\frac{1}{2}a + \sin^2\frac{1}{2}a}.$$

Dividing the numerator and the denominator by $\sin^2\frac{1}{2}a$:

$$\cos a = \frac{\cot^2\frac{1}{2}a - 1}{\cot^2\frac{1}{2}a + 1} = \frac{e^{\frac{2q}{k}} - 1}{e^{\frac{2q}{k}} + 1}.$$

Dividing the numerator and the denominator by $e^{\frac{q}{k}}$:

$$\cos a = \frac{e^{\frac{q}{k}} - e^{-\frac{q}{k}}}{e^{\frac{q}{k}} + e^{-\frac{q}{k}}} = \frac{\sinh\frac{q}{k}}{\cosh\frac{q}{k}} = \tanh\frac{q}{k}. \tag{7}$$

Similarly

$$\sin a = \frac{2\sin\frac{1}{2}a\cos\frac{1}{2}a}{\sin^2\frac{1}{2}a + \cos^2\frac{1}{2}a} = \frac{2\cot\frac{1}{2}a}{\cot^2\frac{1}{2}a + 1} = \frac{2e^{\frac{q}{k}}}{e^{\frac{2q}{k}} + 1}$$

$$= \frac{2}{e^{\frac{q}{k}} + e^{-\frac{q}{k}}} = \frac{1}{\cosh\frac{q}{k}}. \tag{8}$$

We may now write formula (5) for $\smile AK$ (see p. 136) in the form:

$$\smile AK = m\tanh\frac{q}{k}. \tag{9}$$

As an exercise, we recommend the proof of

$$\smile BP = m \sinh \frac{q}{k}. \tag{9'}$$

§ 21. Right-angled triangle

It would not be difficult to deduce by planimetric considerations from the fundamental theorem on the angle of parallelism, proved in the preceding section, the trigonometrical theorems on the right-angled triangle. We shall not, however, take this course, but will examine the stereometric figure which differs from Fig. 99 only in that the vertex C is considered a proper point and not an infinitely distant one.

The description of the figure will only be changed in so far as equality $\beta = \alpha$ is concerned (Fig. 102). This is no longer true, but the other remarks still apply—for intance, that the dihedron with edge k is right and that the dihedron with edge j is equal to α.

As in Fig. 99 we cut the parallel lines j, k, l by a horosphere passing through the point A and get a horosphereic triangle with vertices A, K, L. The angle K of our triangle is right as before, and the angle A is equal to α, so that

$$\smile AK = \smile AL \cos \alpha.$$

The length of $\smile AK$ was given in the last section by formula (9); let us drop the previous notation $q = AB$ and write c instead, as the side of the right-angled triangle ABC. Then

$$\smile AK = m \tanh \frac{c}{k}.$$

The arc AL is obviously as dependent upon the segment $AC = b$ as AK is upon c:

$$\smile AL = m \tanh \frac{b}{k}.$$

We substitute the values for $\smile AK$ and $\smile AL$ in the formula $\smile AK = \smile AL \cos\alpha$:

$$\tanh\frac{c}{k} = \tanh\frac{b}{k}\cos\alpha. \tag{10}$$

We have arrived at the theorem:

The hyperbolic tangent of a side of a right angle (measured in terms of the segment k) *is equal to the product of the hyperbolic tangent of the hypotenuse* (measured in terms of the segment k) and *the cosine of the acute angle adjacent to that side.*

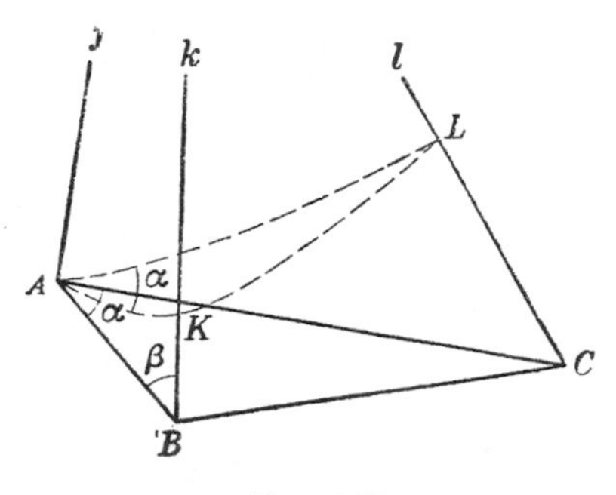

FIG. 102

If we were to draw a horosphere through C instead of A and to use the formula for BP given at the end of the last section, we should obtain the relationship

$$\sinh\frac{a}{k} = \sinh\frac{b}{k}\cdot\sin\alpha, \tag{11}$$

which expresses the theorem:

The hyperbolic sine of the side of a right angle (measured in terms of the segment k) *is equal to the product of the hyperbolic sine of the hypotenuse* (measured in terms of the segment k) *and the sine of the angle opposite to that side.*

The above theorems are the foundation of the trigonometry of right-angled triangles in Lobatchevskian geometry. The former corresponds to the following theorem of ordinary trigonometry: "The side of a right angle is equal to the product of the hypotenuse and the cosine of the acute angle adjacent to that side", and differe from it only in that the hyperbolic tangents of sides (measured in terms of k) appear in place of the sides.

The second theorem differs from the corresponding theorem in Euclidean trigonometry in that sides are replaced by their hyperbolic sines.

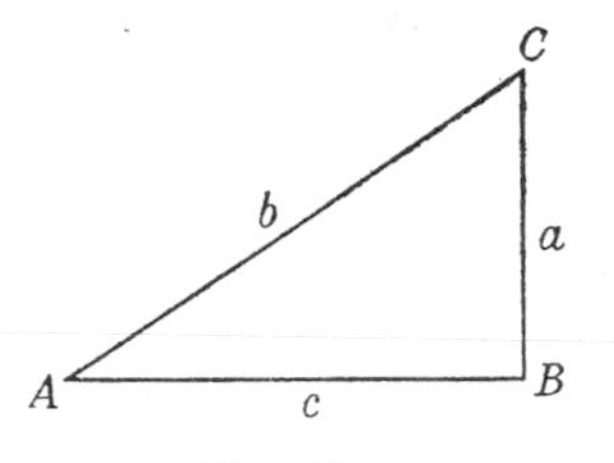

FIG. 103

Of course, the calculations of non-Euclidean trigonometry are more complicated than those of ordinary trigonometry, but by using the tables of hyperbolic functions one may abridge them. This is shown by the following example:

PROBLEM. *In a right-angled triangle, the hypotenuse is* $\frac{1}{2}k$, *and one of the angles is* $45°$. *Find the other two sides and the third angle* (Fig. 103).

We shall measure the sides using k as unit, so that we may omit it as the denominator.

First we apply formula (11)

$$\sinh a = \sinh b \sin 45° = \sinh \tfrac{1}{2} \sin 45°$$

$$= 0{\cdot}521 . 0{\cdot}707 = 0{\cdot}368 .$$

We find from the tables of hyperbolic functions that the value of a is $0{\cdot}360$, therefore

$$a = 0{\cdot}360\, k .$$

Next we apply formula (10)

$$\tanh c = \tanh b \cos 45^\circ = \tanh \tfrac{1}{2} \cos 45^\circ$$
$$= 0{\cdot}462 . 0{\cdot}707 = 0{\cdot}327 .$$

We find from the tables $c = 0{\cdot}340\ k$.

Thus we know all the sides. We find angle C from (11):

$$\sinh c = \sinh b \sin C ,$$

whence

$$\sin C = \frac{\sinh 0{\cdot}340}{\sinh \frac{1}{2}} = \frac{0{\cdot}347}{0{\cdot}521} = 0{\cdot}666 .$$

We find angle C in the tables to be

$$C = 41^\circ 49' .$$

As we see, the methods of solving the right-angled triangle in non-Euclidean geometry involve the same processes as those of Euclidean trigonometry, but demand more searching in tables. Perhaps the greatest inconvenience is the fact that the acute angles of a right-angled triangle do not add up to 90°.

The fundamental theorems do not exhaust the relationships between the elements of the right-angled triangle. They offer equations which connect one side and the hypotenuse with one acute angle, but there must also exist some connection between all three sides or between the two shorter sides and one of the angles. All relationships of this kind may be derived from the fundamental theorems by purely algebraic procedure. As an example we are going to solve the most interesting of the various problems which arise:

Find the equation connecting the hypotenuse b and the sides a and c.

The fundamental theorems give (let us drop, for the moment, the letter k):

$$\sinh a = \sinh b \sin \alpha ,$$
$$\tanh c = \tanh b \cos \alpha .$$

We shall obtain a relationship between the sides of a triangle by eliminating α from the above formulae. To do this we multiply the second equation by $\cosh b$:

$$\tanh c \cosh b = \sinh b \cos \alpha .$$

We square the first and third equations and add them,

$$\sinh^2 a + \tanh^2 c \cosh^2 b = \sinh^2 b .$$

From the identity $\cosh^2 x - \sinh^2 x = 1$ we find $\sinh^2 a$ and $\sinh^2 b$.

$$\cosh^2 a - 1 + \tanh^2 c \cosh^2 b = \cosh^2 b - 1,$$

which gives

$$\cosh^2 a = \cosh^2 b (1 - \tanh^2 c) = \cosh^2 b \, \frac{\cosh^2 c - \sinh^2 c}{\cosh^2 c} .$$

The numerator of the last fraction is 1, whence

$$\cosh^2 a \cosh^2 c = \cosh^2 b ,$$

and remembering that hyperbolic cosines are positive, we have

$$\cosh b = \cosh a \cosh c .$$

Replacing k in the denominators we get the final result

$$\cosh \frac{b}{k} = \cosh \frac{a}{k} \cosh \frac{c}{k} . \tag{12}$$

The hyperbolic cosine of the hypotenuse is equal to the product of the hyperbolic cosines of the other two sides (when the sides and the hypotenuse have been measured in terms of k).

This is the form which the theorem of Pythagoras adopts in Lobatchevskian geometry. There is seemingly no connection between it and its Euclidean version.

Let us take for an illustration $a = 2k$, $c = 2k$:

$$\cosh \frac{b}{k} = \cosh 2 \cosh 2 = 3{\cdot}762 . 3{\cdot}762 = 14{\cdot}15 .$$

We find in the tables

$$\frac{b}{k} = 3{\cdot}342, \quad \text{and so} \quad b = 3{\cdot}342\,k.$$

In Euclidean geometry this would be

$$b = 2{\cdot}828\,k.$$

In a similar manner relationships may be found between other given elements of a right-angled triangle. One of them is the following theorem, which is noteworthy and whose proof is recommended as an exercise to the reader: *the hyperbolic cosine of the hypotenuse measured in terms of k is equal to the product of the cotangents of the acute angles.* It follows that a right-angled triangle is uniquely determined by its acute angles, which confirms the fact known from chapter I that similar triangles do not exist on the Lobatchevskian plane. When, for instance, $\alpha = 30°$, $\gamma = 45°$, a calculation gives $b = 1{\cdot}15\,k$. For smaller angles the hypotenuse b would assume greater values.

§ 22. The geometry of sufficiently small domains

The theorems on right-angled triangles in the geometries of Euclid and of Lobatchevsky are different, and so the results of solving triangles are also different. If the shorter sides are a and c, we get, in Euclidean geometry,

$$b = \sqrt{a^2 + c^2},$$

whereas in Lobatchevskian geometry b is determined by the formula (12) of the last section:

$$\cosh\frac{b}{k} = \cosh\frac{a}{k}\cosh\frac{c}{k}.$$

If we substitute

$$\cosh\frac{b}{k} = \frac{e^{\frac{b}{k}} + e^{-\frac{b}{k}}}{2},$$

and then determine $e^{\frac{b}{k}}$ from the equation obtained, we shall finally have

$$\frac{b}{k} = \log_e\left(\cosh\frac{a}{k}\cosh\frac{c}{k} + \sqrt{\cosh^2\frac{a}{k}\cosh^2\frac{c}{k} - 1}\right);$$

this formula resembles in no way the classical root of the sum of squares. The difference in analytical form, however, does not necessarily mean that the two formulae will not give similar results. Therefore it would be natural to examine this difference and discuss whether it is small or large. It appears that the difference is slight if sides a, b, c are small compared with k, i. e. if the ratios $\frac{a}{k}, \frac{b}{k}, \frac{c}{k}$ are sufficiently small.

In this case $\cosh y$ may be replaced in the formulae by $1+\frac{1}{2}y^2$, producing an error of λy^4, where λ is positive and, according to the theorem of § 19, less than $\frac{1}{20}$. We do not need closer information about λ, we know only that it depends on y.

The formula (12) will now look like

$$1+\frac{b^2}{2k^2}+\lambda\frac{b^4}{k^4} = \left(1+\frac{a^2}{2k^2}+\lambda_1\frac{a^4}{k^4}\right)\left(1+\frac{c^2}{2k^2}+\lambda_2\frac{c^4}{k^4}\right).$$

Multiplying out the terms in brackets, deleting the unit, putting $\lambda\frac{b^4}{k^4}$ on the right-hand side and finally multiplying both sides by $2k^2$, we get

$$b^2 = a^2+c^2+\frac{4\lambda_1 a^4+4\lambda_2 c^4+a^2c^2-4\lambda b^4}{2k^2}+$$

$$+\frac{\lambda_1 a^4c^2+\lambda_2 a^2c^4}{k^4}+\frac{2\lambda_1\lambda_2 a^4c^4}{k^6}.$$

This shows that b^2 differs from a^2+c^2 by the last three terms, the sum of which shall be denoted by R.

We shall estimate R approximately by replacing all the segments contained in it by b, the greatest of them, replacing $-4\lambda b^4$ by $+4\lambda b^4$, and finally all the λ's by $\frac{1}{20}$:

$$|R| < \frac{4}{5}\cdot\frac{b^4}{k^2}+\frac{1}{10}\cdot\frac{b^6}{k^4}+\frac{1}{200}\cdot\frac{b^8}{k^6}.$$

Thus b^2 differs from a^2+c^2 by a number whose ratio to b^2 is less than $\frac{R}{b^2}$, i. e. than

$$w = \frac{4}{5}\cdot\frac{b^2}{k^2}+\frac{1}{10}\cdot\frac{b^4}{k^4}+\frac{1}{200}\cdot\frac{b^6}{k^6}$$

$$= \frac{4}{5}\cdot\frac{b^2}{k^2}+\frac{1}{5}\cdot\frac{b^2}{k^2}\left(\frac{1}{2}\cdot\frac{b^2}{k^2}+\frac{1}{40}\cdot\frac{b^4}{k^4}\right).$$

If $\frac{b}{k}$ is less than 1, then the number in brackets is less than 1, whence

$$w < \frac{b^2}{k^2} = \left(\frac{b}{k}\right)^2.$$

This means that the difference between the square of the hypotenuse b, calculated by Lobatchevskian geometry, and the sum of the squares of the shorter sides does not exceed a number whose ratio to b^2 is less than $\left(\frac{b}{k}\right)^2$, provided that the hypotenuse is less than the segment k.

This ratio is known as the relative error. Let us take, as a numerical illustration, a right-angled triangle with sides less than a hundred-thousandth part of the segment k. Then, when calculating b^2 as the sum of the squares of the shorter sides, we shall obtain a result differing from the correct value by at most $(0{\cdot}00001)^2 = \frac{1}{10^{10}}$ th part of it. It is evident that the error is so small that it cannot be noticed in empirical measurements. For triangles whose sides are smaller it is the more true.

In sufficiently small domains of Lobatchevskian space the theorem of Pythagoras in its Euclidean form gives practically correct results. "Sufficiently small domain" means if its diameter is sufficiently small compared to k, for instance, if its ratio to k is less than 0·00001.

The same applies to other theorems, e. g. to the theorem on the sum of the angles of the triangle.

For simplicity's sake let us consider a right-angled triangle with sides a, b, c and determine the sines and cosines of the angles by the fundamental theorems. Again, we omit for the time being the figure k in the denominators:

$$\sin\alpha = \frac{\sinh a}{\sinh b}, \qquad \cos\alpha = \frac{\tanh c}{\tanh b},$$

$$\sin\gamma = \frac{\sinh c}{\sinh b}, \qquad \cos\gamma = \frac{\tanh a}{\tanh b}.$$

The defect δ of the triangle is equal to $90^\circ - \alpha - \gamma$. Let us evaluate its sine

$$\sin\delta = \sin(90^\circ - \alpha - \gamma) = \cos(\alpha + \gamma)$$
$$= \cos\alpha\cos\gamma - \sin\alpha\sin\gamma.$$

Let us substitute the given values of $\sin\alpha$, $\cos\alpha$, $\sin\gamma$, $\cos\gamma$, when $\tanh b$ has been replaced by $\dfrac{\sinh b}{\cosh b}$. We get

$$\sin\delta = \frac{1}{\sinh^2 b}(\tanh a \tanh c \cosh^2 b - \sinh a \sinh c).$$

For one $\cosh b$ let us substitute $\cosh a \cosh c$ according to the formula (12):

$$\sin\delta = \frac{1}{\sinh^2 b}(\sinh a \sinh c \cosh b - \sinh a \sinh c)$$
$$= \frac{\sinh a \sinh c(\cosh b - 1)}{\sinh^2 b}.$$

But $\sinh^2 b = \cosh^2 b - 1 = (\cosh b - 1)(\cosh b + 1)$, therefore

$$\sin\delta = \frac{\sinh a \sinh c}{1+\cosh b} = \frac{\sinh a \sinh c}{1+\cosh a \cosh c}.$$

Replacing k in the denominators,

$$\sin\delta = \frac{\sinh\frac{a}{k}\,\sinh\frac{c}{k}}{1+\cosh\frac{a}{k}\,\cosh\frac{c}{k}}. \tag{13}$$

We have expressed the defect of the triangle as a function of its shorter sides.

Using the approximate formulae of § 19 we shall obtain, on the assumption $a < k$, $c < k$, an approximate value of $\sin\delta$. We shall take only the first terms of these formulae and drop those whose ratio to the former is of the order y^2:

$$\sin\delta = \frac{\frac{a}{k}\cdot\frac{c}{k}}{1+1.1} = \frac{ac}{2k^2}.$$

The neglected terms would have k^4 in the denominator.

An easy calculation would give the more accurate value:

$$\sin\delta = \frac{ac}{2k^2}\left(1-\frac{a^2+c^2}{12k^2}\right),$$

but the first will suffice for our purposes.

It will be convenient to use further the radian-measure of angles. The radian-measure of angle α is $\frac{\pi\alpha}{180}$. We have the following theorem: *the sine of a sufficiently small angle has nearly the same value as the radian-measure of this angle.* More precisely, *the ratio of the sine of an angle to its radian-measure tends to unity as the angle tends to zero*. Assuming that the defect δ of a triangle has been expressed in radian-measure, we obtain the approximate formula

$$\delta = \frac{ac}{2k^2}.$$

More precisely, the limit of $\frac{\delta}{ac}$, as a and c tend to zero, is equal to $\frac{1}{2k^2}$.

If $\frac{a}{k}$ and $\frac{c}{k}$ were less than, say, 0·0001, the defect would be less than $\frac{1}{200,000,000}$.

One second in radians is $\frac{48}{10,000,000}$; our defect is 1000 times smaller, that is, it is less than 0·001″. In empirical measurements such an angle cannot be detected, so that the sum of the angles of a triangle in sufficiently small domains does not differ significantly from 180°.

Summarizing, *in sufficiently small domains of Lobatchevskian space Euclidean geometry is in force*. What this means is clear from what has been said.

Let us apply the formula to settling the question we were concerned with on p. 89. There it was stated that the area of a polygon in Lobatchevskian geometry may be defined as the product of its defect and a certain constant λ, the value of which had not been determined

$$P = \lambda\delta.$$

Let us now return to this, and consider a right-angled triangle with short sides a, c, area P and defect δ. It follows from the above that

$$\frac{P}{\frac{1}{2}ac} = \lambda\frac{\delta}{\frac{1}{2}ac}.$$

If the sides a and c tend to zero, the ratio $\frac{\delta}{\frac{1}{2}ac}$ will tend to $\frac{1}{k^2}$ owing to our previous remarks, whence $\frac{P}{\frac{1}{2}ac}$ will tend to $\lambda \frac{1}{k^2}$.

For sufficiently small a and c Euclidean geometry holds good, according to which the area of a right-angled triangle is given by the formula $P = \frac{1}{2}ac$. So it follows that for the sake of consistency of the definition of area in Lobatchevskian space with the concepts of the Euclidean geometry which holds in sufficiently small domains of it we ought to take $\lambda = k^2$. Therefore, *the area of a polygon is equal to the product of its defect and the square of the constant* k.

The defect of a triangle does not exceed 180°, or π in radians, so no triangle may have an area greater than πk^2. Similarly, the area of a square, and more generally that of a quadrangle, is always less than $2\pi k^2$.

Let us consider still the defects of right-angled triangles whose shorter sides vary in length between $\frac{1}{1000}k$ and $\frac{1}{100}k$. If $a = c = \frac{1}{1000}k$, then $P = \frac{1}{2{,}000{,}000}k^2$ and $\delta = \frac{P}{k^2} = \frac{1}{2{,}000{,}000}$, which gives $\frac{1}{10}$ sec. when converted from radians into degrees. If $a = c = \frac{1}{100}k$, then $\delta = 10''$.

The defects of triangles with sides less than $\frac{1}{1000}k$ are unobservable. If we succeeded in finding a triangle with area P (evaluated in the Euclidean manner) and in measuring its defect δ we might thereby determine the constant k from the formula

$$k^2 = \frac{P}{\delta}.$$

§ 23. Lobatchevskian geometry and the empirical space

So far we have been discussing theoretical matters. We shall now apply them to the study of empirical space and the geometrical properties of it by asking once more the fundamental question: does Lobatchevskian or Euclidean geometry govern the real space? The matter would be settled if we could find a triangle whose defect was not zero. In addition, the constant k would also be determined thereby, in accordance with the remark made at the end of the last section. Our knowledge of the universe would take an all-important step forward; astronomy would be most affected, and many of its theories would have to be modified.

As was said on p. 52, no triangle with an observable defect has yet been found. However, we are not entitled to maintain that Euclidean geometry holds good throughout all space. Our failure might simply be due to the fact that the areas of all triangles accessible to us are too small when compared with the quantity k^2, as was indicated in § 22, or in other words, that the constant k is too great in comparison with the figures of our astronomy. We will support this somewhat vague observation with some argument.

We showed on p. 51 that the triangle, one side of which is the radius of the earth's orbit round the sun and whose opposite vertex is at the fixed star G, has a defect which is, to be sure, unknown but which is certainly less than the parallax of the star—an angle which has been measured for some stars, and which for Sirius is 0·38″. Thus it follows that the right-angled triangle AGM (Fig. 30), one of whose shorter sides is the distance from the earth to the sun (93,000,000 miles) and the other the distance from the earth to Sirius (about 50 trillion miles), has a defect of less than 0·38″, or $\frac{18}{10,000,000}$ radians. Suppose that the defect δ of a triangle is less than a certain angle α.

It follows from the formula $\delta = \frac{P}{k^2}$, where P stands for the area of $\triangle AGM$, that

$$\frac{P}{k^2} < \alpha$$

which gives

$$k > \sqrt{\frac{P}{a}}.$$

On substituting the known values for Sirius and other stars we should elicit that the constant k is not less than about six hundred trillion miles. Therefore the measurement of the parallaxes of fixed stars gives results which are consistent both with the conjecture that Euclidean geometry holds in space and with that that Lobatchevskian geometry holds, when the constant k is greater than six hundred trillion miles.

We ought not to expect that purely geometrical measurements would decide between these possibilities, nor, if the latter case was closer to reality, that they would enable us to estimate k. Perhaps it might be done one day by some other method, as we mentioned on p. 52. The theory of relativity has formulated a hypothesis that the value of k is connected with the physical phenomena occurring in space. Therefore the observation of these phenomena might lead to the value of k and decide whether it is finite or not. If not, we would have simply the geometry of Euclid. So far the long and wearisome work that would have to go into this has not been carried out but, according to Einstein, it is within the powers of modern science.

Here we may make room to present and dismiss some of the philosophico-physical objections which were promulgated in the early days of non-Euclidean geometry, and which were directed against the existence in space of the length k, the natural "unit of length". In the general sense, any segment may be taken as a unit of measure, all segments having equal rights in this respect. Gauss made the following comment in his correspondence: "The only thing in this science which is opposed to our reason is that if this science were true, there must exist in space a certain length determined in itself (although unknown to us). But, methinks, in spite of the meaningless word-wisdom of our metaphysicians, we know too little or nothing at all about the real meaning of space to stamp anything appearing unnatural to us as absolutely impossible."

This opinion, which bears witness to the uncommon independence of Gauss's views, gives no reason why the human mind should have recoiled from the hypothesis of the existence of an absolute unit of length. The reason for this is to be sought in the philosophical speculations which regarded space as entirely separated from the matter it contained. Space used to be considered as, so to say, a frame into which a picture could later be inserted, or as a vessel into which milk might be poured but which existed independently of the milk. According to this view space was homogeneous and no segment was distinguishable in it.

"The concept of such space", declared Kant (1), "is by no means of empirical origin", but is an inevitable necessity of thought and does not depend upon the existence of any objects, since "it is impossible to imagine that there is no space but it is possible, without any difficulty, to imagine that there are no objects in it". To Kant, geometry was an "apodictic science", that is, a science which is forced upon us by our consciousness.

Gauss's caustic taunt about the empty verbiage of the metaphysicians was certainly directed against the quoted opinions of Kant, for Gauss, as we have seen, considered observation to be the origin of geometrical knowledge and denied importance to the question of whether or not we can imagine anything.

The existence of a length defined by itself is contrary not so much to the reason as to the habit of forming concepts in which space has been dissociated from the matter which occupies it. Contemporary physical theories

---

(1) In the *Critique of Pure Reason*, 1781.

oppose these concepts and state that the geometrical properties of space depend on the distribution of the matter in it, and are different in neighbourhoods where it is concentrated from those where it is not. The structure of these theories is complicated, but we shall give an outline of their principal ideas if we explain that in the neighbourhood of greater masses non-Euclidean geometry holds good approximately, but with a value of k which is different from that in the neighbourhood of smaller masses. Thus the geometry of real space is so to speak modelled by matter and the study of space cannot be divorced from the study of matter. Euclidean geometry, as well as non-Euclidean with k constant throughout space, is only a simplified picture of reality which is sufficient for most practical applications but cannot claim to give the whole truth about the universe.

These ideas, some beginnings of which may be traced in the lecture of Riemann mentioned in § 7, were formulated by Albert Einstein in an imposing scientific theory, in which the science of the geometrical properties of real space was connected with the science of the properties of matter; geometry has thereby become a department of physics, of empirical science. Yet, the opinion that geometry cannot be treated as an *a priori* science, a product of some necessity of thought, but as a science resting on empirical foundations, is earlier than the theory of relativity. The creators of non-Euclidean geometry must have entertained it, for it is closely bound up with this geometry. After all, it is possible to assert that the axiom according to which only one straight line passes through two points is the fruit of pure reason, but it would obviously be quite absurd to wish to discover by meditation the length in miles of the constant k. Therefore the birth of non-Euclidean geometry struck a mortal blow at the idealistic conceptions of Kant which were ruling at the beginning of the last century and which tolerated no opposition. Gauss was in no small apprehension of disclosing his ideas and wrote to his friends that that would have caused a "howl from the Boeotians" (1).

We have given a brief account of the philosophical implications of the discoveries of Lobatchevsky and Bolyai, and will now return to purely geometrical topics, and namely to various applications of the theorems on the right-angled triangle. We shall not trouble about giving a full description of these applications nor about systematizing them; we shall merely present a few interesting questions.

§ 24. Quadrangle with three right angles

The quadrangle of Saccheri and the half of it, the quadrangle with three right angles, are figures which are constructed in a simple manner. We construct a quadrangle with three right angles as follows: we draw perpendiculars AD and BC to the segment AB (Fig. 104), choose a point C on the latter perpendicular, and drop from it a perpendicular CD to the straight line AD. The figure is fully determined by the two segments AB and BC which may be of any length. Three angles of our quadrangle are right, the fourth C is acute. The straight line AC would split the quadrangle into two right-angled triangles, and various problems concerned with the quadrangle may be reduced to the examination of these two triangles.

The most interesting of them refers to the property which was, on p. 97, used in the construction of a line parallel to a given line. We may state it as follows:

Evaluate the angle between the side AD of the quadrangle and the segment AK, where $AK = CD$.

Let us write $AB = a$, $CD = b$, $AD = c$; we will drop k in the denominators, i. e. take k as our unit. We deduce

(1) In ancient Greece the Boeotians were considered a crude people.

from the fundamental theorem (p. 143) applied to $\triangle KAB$

$$\cos(90^\circ - \alpha) = \sin\alpha = \frac{\tanh a}{\tanh b}. \tag{14}$$

On the other hand we get, similarly, for $\triangle ABC$

$$\tanh a = \tanh AC \cos\beta,$$

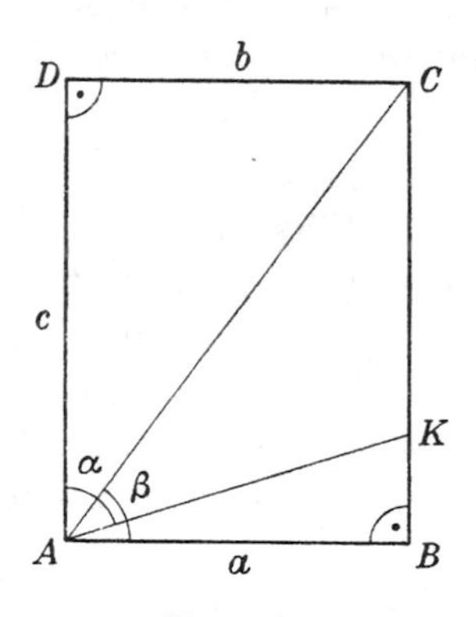

FIG. 104

and applying the fundamental theorem (11) (p. 143) to $\triangle ACD$

$$\sin \sphericalangle CAD = \sin(90^\circ - \beta) = \frac{\sinh b}{\sinh AC} = \cos\beta.$$

Substituting this value of $\cos\beta$ into the preceding formula we get

$$\tanh a = \tanh AC \frac{\sinh b}{\sinh AC} = \frac{\sinh b}{\cosh AC}.$$

But AC is the hypotenuse of the triangle ACD, whence by (12), p. 146

$$\cosh AC = \cosh b \cosh c,$$

$$\tanh a = \frac{\sinh b}{\cosh b \cosh c} = \frac{\tanh b}{\cosh c}$$

and finally substituting the value $\tanh a$ in (14) we arrive at the formula

$$\sin \alpha = \frac{1}{\cosh c}. \tag{15}$$

This result is certainly unexpected. It appears that the acute angle α depends only on the side c of the quadrangle and is independent of the side a. Comparing (15) with the formula (8) $\sin \alpha = \dfrac{1}{\cosh q}$, we notice that the rôle of q is now assumed by c. The formula (8) gives the angle of parallelism corresponding to the distance q, hence the formula (15) gives the angle of parallelism corresponding to the distance c, which means that AK is parallel to b. Bolyai's construction of a line parallel to a given line through a point outside it was based on precisely this observation.

Bolyai's construction needs the use of a ruler and a pair of compasses. In non-Euclidean geometry there appear, in addition to the circle, other circular lines, the equidistant and the horocycle, which may also slide along themselves. It would not be difficult to devise instruments to draw these mechanically. Thus, in addition to constructions involving straight lines and circles, one may think of constructions performed by means of these new drawing-instruments, by, say, an ordinary ruler and a horocyclic ruler. Consequently, the constructional methods are more varied in non-Euclidean geometry than in Euclidean. It may, however, be shown that each construction that uses all four instruments could also be carried out by means of ruler and compasses, though in many cases in a more complicated manner. But not every construction possible on the Euclidean plane can be carried out on the Lobatchevskian plane, for some of the former make use of the properties of parallel lines which do not hold in Lobatchevskian geometry. Such a Euclidean construction is the division of a segment into n ($n \neq 2^k$) equal parts, which depends on the theorem that if parallel straight lines cut off equal segments on one side of an angle they will also cut off equal segments on the other side. This theorem is not true in Lobatchevskian geometry and the construction of dividing a segment into n equal parts also fails. A closer examination shows that it is not possible, using only ruler and compasses, to divide an arbitrary segment even into three equal parts.

We do not propose to study these questions more closely, since we have not included the extensive theory of construction problems (apart from the construction of Bolyai) in the present book of limited size.

The "map" of section 10 has proved useful in a number of construction problems. With its help we shall solve, as an example, the following classical problem.

To find a segment q, the angle of parallelism α corresponding to which is given.

Let α be the angle of parallelism for point O and straight line a (Fig. 105a). We choose O as the centre of the map and segment r as its radius. This means that the angle β of transformation (see p. 70) has been chosen so that the projection of the whole of one side of it onto the other is a segment OH_0 of length r (Fig. 105b). In other words we construct the angle β by taking $OH_0 = r$ and drawing first the straight line b perpendicular to it at the point H_0 and second the line parallel to b through O, which we are able to do (Bolyai's construction).

Fig. 105a will be represented on our map by Fig. 105c, where the angle $C'OD' = \alpha$, OC' represents the wanted segment and is equal to a short side of the triangle with hypotenuse $OD' = r$ and known angle α. The segment OC' may be easily constructed. We place it on the side OH_0 of the angle β (Fig. 105b) and find the segment OC at once.

Another interesting problem concerning the quadrangle $ABCD$ (Fig. 104) with three right angles is connected with the question: Can we obtain such a quadrangle by choosing two arbitrary points B and D on the sides of a right angle A and drawing perpendiculars at these points to the sides ot the angle?

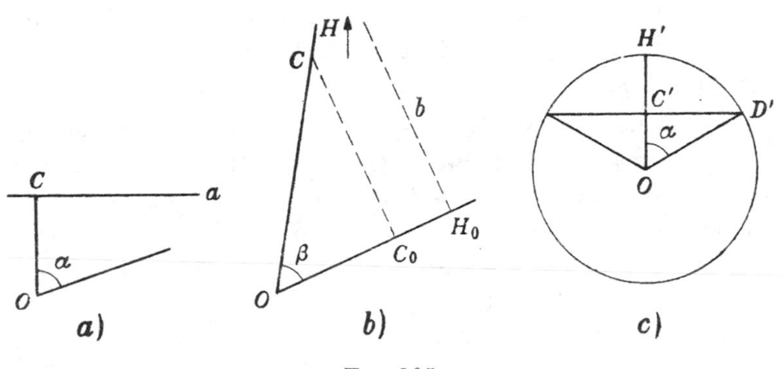

FIG. 105

Obviously the answer is yes, under the condition, however, that these perpendiculars will intersect. Therefore we put the question:

What are necessary and sufficient conditions that a perpendicular to one side of a right angle cut the perpendicular to the other side?

On Fig. 104 the perpendiculars to the sides of the right angle A cut them at the points B and D. Let us write $AB = a$, $AD = c$.

We are seeking the conditions for the existence inside the angle A of a point C whose perpendicular projections onto the sides of the angle would be the points B and D.

In other words a segment $AC = r$ should exist which would form an acute angle β with AB and have perpendicular projections a and c onto the sides of the right angle respectively. These projections may be evaluated by using the fundamental theorems on right-angled triangles

(p. 143). We obtain the equations

$$\begin{aligned} \tanh a &= \tanh r \cos\beta, \\ \tanh c &= \tanh r \cos(90^\circ - \beta) = \tanh r \sin\beta. \end{aligned} \tag{16}$$

Therefore, the question is whether there exist numbers r and β which satisfy these equations. We square them and add:

$$\tanh^2 a + \tanh^2 c = \tanh^2 r.$$

Now, let us recall that the hyperbolic tangent is a function which assumes all values between 0 and 1, excluding 1. Therefore, in order that segment r exist, it is necessary and sufficient that

$$\tanh^2 a + \tanh^2 c < 1. \tag{17}$$

If r exists, we shall find β directly from the formulae (16). Therefore, in order that our perpendiculars intersect it is necessary and sufficient that the inequality (17) hold.

It is easy to see that in the case where

$$\tanh^2 a + \tanh^2 c = 1,$$

the perpendiculars to both sides of the angle will be parallel, and if

$$\tanh^2 a + \tanh^2 c > 1,$$

they will be divergent straight lines.

§ 25. Various problems

PROBLEM. *Find the side of a regular n-sided polygon inscribed in a circle with radius r.*

Let $ABC\ldots$ be a regular n-sided polygon (Fig. 106). In the triangle AOL the side $OA = r$, $\sphericalangle AOL = \alpha = \dfrac{180^\circ}{n}$, $AB = 2a$.

Then $\sinh \frac{a}{k} = \sinh \frac{r}{k} \sin \alpha$ by the theorem on right-angled triangles (see (11), p. 143), whence the segment a has been determined.

PROBLEM. *Calculate the radius of a circle inscribed in a regular n-sided polygon of side a.*

In Fig. 106 we are given the segment $AB = 2a$, $\sphericalangle AOL = \alpha = \frac{180°}{n}$, $OA = r$ and the problem is reduced to finding the side $OL = \varrho$ of the triangle OAL in which the other short side $AL = a$ and the angle α opposite it are known.

It follows from the theorem on triangles ((11), p. 143) that

$$\sinh OA = \frac{\sinh a}{\sin \alpha},$$

and from another theorem ((10), p. 143) that

$$\tanh \varrho = \cos \alpha \tanh OA = \frac{\cos \alpha \sinh OA}{\cosh OA}.$$

In these transformations we have dropped k in the denominators.

Substituting the value $\sinh OA$ in the numerator and bearing in mind that (formula (12), p. 146) $\cosh OA = \cosh \varrho \cosh a$, we get

$$\tanh \varrho = \frac{\cot \alpha \sinh a}{\cosh a \cosh \varrho},$$

which gives when multiplied by $\cosh \varrho$

$$\sinh \varrho = \cot \alpha \tanh a.$$

This formula leads to an interesting corollary. Let the side a of a regular n-sided polygon tend to infinity. In this case (p. 131) $\tanh a$ will tend to unity, whence $\sinh \varrho$ will tend to $\cot \alpha$, and the radius of the circle inscribed in the regular n-sided polygon will therefore not exceed the value ϱ_0 given by the equation

$$\sinh \varrho_0 = \cot \frac{180°}{n}.$$

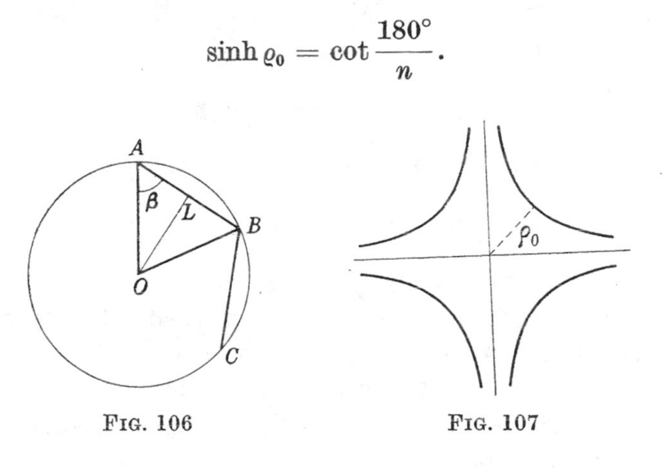

FIG. 106 FIG. 107

In the case $n = 4$ we have $\sinh \varrho_0 = 1$ and $\varrho_0 = 0{\cdot}881\ldots$, and replacing k, the distance of the centre of a square from its side is always less than $k \,.\, 0{\cdot}881\ldots$

To the value $\varrho = \varrho_0$ there corresponds the square with infinitely distant vertices whose sides are lines parallel to two mutually perpendicular straight lines (Fig. 107). This square may be considered as the "biggest" possible square. Its area is $2\pi k^2$ (§ 22).

PROBLEM. *Find the angle of a regular n-sided polygon inscribed in a circle of radius r.*

We proceed to find $\sphericalangle OAL = \beta$ in the triangle OAL (Fig. 106). By the theorem on right-angled triangles (p. 143) we have

$$\cos \beta = \frac{\tanh a}{\tanh r}.$$

Substituting the value $\tanh a$ from the last problem

$$\cos \beta = \frac{\sinh \varrho \tan \alpha}{\tanh r} = \frac{\sinh \varrho \tan \alpha \cosh r}{\sinh r}.$$

By the theorem on triangles ((11), p. 143) we get

$$\frac{\sinh \varrho}{\sinh r} = \sin \beta,$$

whence

$$\cos \beta = \sin \beta \tan \alpha \cosh r.$$

Dividing both sides by $\sin \beta$

$$\cot \beta = \tan \alpha \cosh r.$$

This formula solves the problem.

So far our problems have been concerned with the theory of regular polygons. We shall now discuss the most important question about arbitrary triangles:

Given two sides of a triangle and the angle between them find the third side.

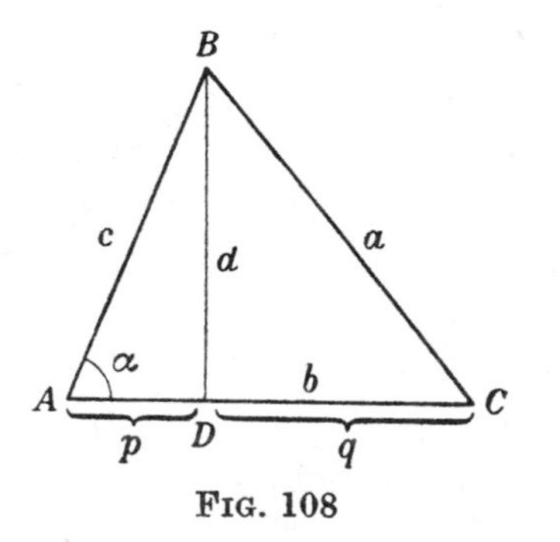

FIG. 108

Let us take the triangle ABC (Fig. 108). Let us draw the altitude BD from the apex B. We shall discuss in particular the case where our altitude cuts the opposite side. If this were not so, we should have to change the signs in some of the formulae but the final result would remain the same. The reader is recommended to verify this.

The working given below is somewhat longer than in the analogous Euclidean problem but both calculations rest on the same basic ideas: the application of the theorem of Pythagoras to triangle BCD and of both fundamental theorems to triangle ABD.

By the theorem of Pythagoras we have in triangle BCD

$$\cosh a = \cosh q \cosh d.$$

Segment $q = b-p$, so, as on p. 131, we have

$$\cosh q = \cosh b \cosh p - \sinh b \sinh p.$$

Therefore

$$\cosh a = \cosh b \cosh p \cosh d - \sinh b \sinh p \cosh d.$$

Turning to triangle ABD we apply the theorem of Pythagoras and get

$$\cosh p \cosh d = \cosh c, \tag{18}$$

then we replace $\sinh p$ by $\tanh p \cosh p$ in the formula for $\cosh a$ and we deduce

$$\cosh a = \cosh b \cosh c - \sinh b \cosh p \tanh p \cosh d.$$

The theorem on triangles (p. 143) applied to triangle ABD gives

$$\tanh p = \tanh c \cos \alpha.$$

Substituting this value in the second term of the formula for $\cosh a$ we write it as

$$\sinh b \cosh p \tanh c \cos \alpha \cosh d,$$

and taking into account (18) we obtain

$$\sinh b \tanh c \cosh c \cos \alpha = \sinh b \sinh c \cos \alpha.$$

Finally, the third side of our triangle will be determined by the formula

$$\cosh a = \cosh b \cosh c - \sinh b \sinh c \cos \alpha. \tag{19}$$

The corresponding formula in Euclidean geometry is

$$a^2 = b^2 + c^2 - 2bc \cos \alpha.$$

We call these formulae cosine theorems in Euclidean trigonometry and in the (hyperbolic) trigonometry of Lobatchevsky. Their applications are similar, namely, they enable one not only to find the side of a triangle when given the two remaining sides and the angle between them, but also to find the angle when given all the sides. This initial resemblance implies that in their further development the two trigonometries, which differ as far as the appearance of their formulae is concerned, are quite close to each other in their basic ideas and methods. We shall not discuss this any more closely as this would exceed the scope of the present book.

§ 26. Length of a circle and other problems

The length of the circumference of a circle may be defined as the limit of the length of the perimeter of a regular polygon inscribed in the circle as the number of sides of the polygon tends to infinity.

In order to find the length of an arbitrary circumference we shall use the following theorem: *If the radius r of a circle is sufficiently small, the length l of the circumference is determined by the laws of Euclidean geometry*, i. e. it is approximately equal to $2\pi r$; more precisely, the ratio $\frac{l}{2\pi r}$ tends to unity as r tends to zero. This theorem is a particular case of the more general one that *in sufficiently small domains of the Lobatchevskian plane Euclidean geometry is in force.* We have discussed this principle in detail for the theorem of Pythagoras and for the theorem on the sum of the angles of a triangle. It would not be difficult to prove that it applies also to other theorems.

Let us now consider two concentric circles of radii R and r respectively. Let us inscribe in each of them a regular n-sided polygon and denote the perimeters of the polygons by P_n and p_n. According to the first problem of the last section the perimeters of these polygons are to one another as $\sinh\frac{R}{k} : \sinh\frac{r}{k}$, or

$$\frac{P_n}{p_n} = \frac{\sinh\frac{R}{k}}{\sinh\frac{r}{k}}.$$

As n tends to infinity P_n and p_n tend to the circumferences L and l of the two circles, whence

$$\frac{L}{l} = \frac{\sinh\frac{R}{k}}{\sinh\frac{r}{k}}, \qquad L = \sinh\frac{R}{k}\cdot\frac{l}{\sinh\frac{r}{k}},$$

which may be written:

$$L = \sinh\frac{R}{k}\cdot\frac{l}{2\pi r}\cdot 2\pi k\cdot\frac{\left(\frac{r}{k}\right)}{\sinh\frac{r}{k}}.$$

By the theorem at the end of § 19 the latter factor tends to 1 as r, and consequently $\frac{r}{k}$, tends to zero. On the other hand the fraction $\frac{l}{2\pi r}$ also tends to unity (see above). Therefore the right-hand side of the formula tends to $\sinh\frac{R}{k}\cdot 2\pi k$, and so the left-hand side tends to the same limit, but since it is equal to L we get

$$L = 2\pi k \sinh\frac{R}{k}. \tag{20}$$

We have calculated the length of the circumference of a circle of radius R. Its approximate value for small

$\frac{R}{k}$ (p. 133) is

$$L = 2\pi k\left(\frac{R}{k} + \frac{1}{6}\cdot\frac{R^3}{k^3}\right) = 2\pi R + \frac{\pi R^3}{3k^2}$$

so it is greater than that of Euclidean geometry and becomes much greater still if $R > k$, e. g. for $R = 5k$ the circumference of the non-Euclidean circle is nearly fifteen times greater than $2\pi R$.

A similar argument applies to the area of a sphere. We need first to give a precise definition of that area. It is usually done by inscribing cones in the sphere and considering the area in question as the limit of the areas of these cones as the cones change in accordance with a certain convention. We do not propose carrying out an analysis of this sort but will content ourselves with a summary presentation. One may, however, transform it into a complete proof. Let us imagine two concentric spheres of radii R and r (Fig. 109).

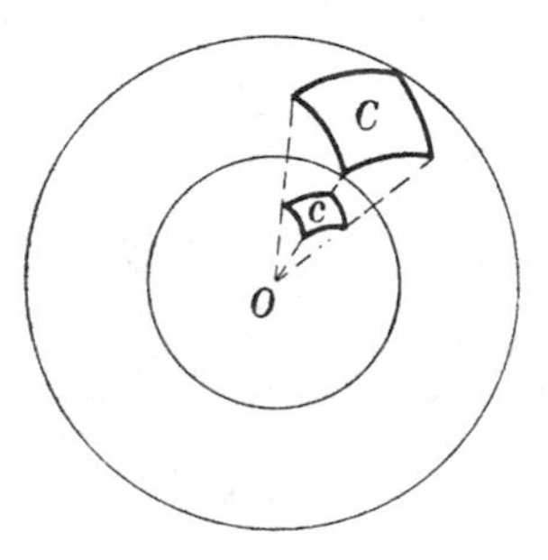

FIG. 109

On the surface of the larger sphere let us consider a small quadrangle C formed, say, by the arcs of great circles. Drawing radii through the points on the perimeter of C and letting them intersect the surface of the smaller sphere, we shall obtain the quadrangle c on this surface. The sides of both quadrangles are arcs of circles with radii R and r respectively, so from the last problem they are to one another as $\sinh\frac{R}{k} : \sinh\frac{r}{k}$.

If the quadrangles are small enough Euclidean geometry works and the areas of the quadrangles are proportional to squares of corresponding sides, i. e.

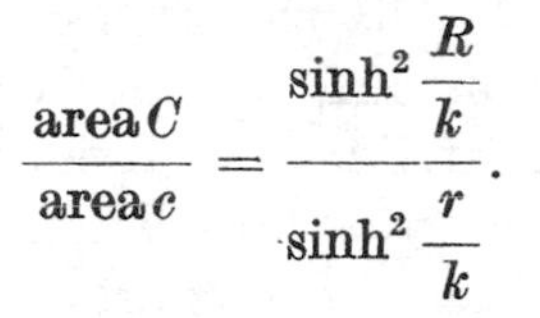

$$\frac{\text{area}\,C}{\text{area}\,c} = \frac{\sinh^2\frac{R}{k}}{\sinh^2\frac{r}{k}}.$$

The areas P and p of the spheres are in the same proportion since both surfaces may be split up into a number of quadrangles C and c:

$$\frac{P}{p} = \frac{\sinh^2\frac{R}{k}}{\sinh^2\frac{r}{k}}, \quad \text{whence} \quad P = p\frac{\sinh^2\frac{R}{k}}{\sinh^2\frac{r}{k}}.$$

We then proceed exactly as with the circumference of the circle. We write the latter formula as

$$P = \frac{p}{4\pi r^2}\cdot\sinh^2\frac{R}{k}\cdot 4\pi k^2\cdot\frac{\left(\frac{r}{k}\right)^2}{\sinh^2\frac{r}{k}},$$

and find the limit as r tends to zero. The factor $\frac{p}{4\pi r^2}$ tends to 1, since Euclidean geometry holds in small domains, the factor $\frac{\left(\frac{r}{k}\right)^2}{\sinh^2\frac{r}{k}}$ also tends to 1, and finally

$$P = 4\pi k^2\sinh^2\frac{R}{k}. \qquad (21)$$

We have evaluated the area of the surface of a sphere of radius R.

Observations similar to those made when discussing the circumference of a circle apply to formula (21). The area of a sphere of radius R is greater in Lobatchevskian geometry than in Euclidean, and for $R > k$ it is even much greater. If $R = 5k$ the area of the sphere is about 220 times as great as $4\pi R^2$, which is its Euclidean value.

Of course, such striking deviations from the laws of Euclidean geometry would occur in figures whose sizes were of the same order as the constant length k or greater. The reader might therefore easily conclude that the study of the properties of such enormous objects, though interesting in theory, can be of no practical purpose, for objects of this size are not accessible to man. This is certainly true in most sciences, but modern astronomy has detected the existence of celestial objects, other galactic systems, which are hundreds of thousands and millions of light-years away from us, and has entered with incomparable skill upon the task of discovering the mutual positions and distances of these objects. The measurements and calculations of the astronomers, rewarded with splendid success, have rested on the assumption that Euclidean geometry governs the universe. If it turned out that the geometrical properties of the universe were better described by the science of Lobatchevsky with a certain "average" value of k, the results of astronomy so far obtained would have to be corrected, perhaps even to a considerable extent. In this case the considerations of non-Euclidean geometry would prove of great significance to astronomy and would exercise considerable influence on our views about the structure of the universe.

We shall discuss this more closely. The most important method of estimating the distances of spiral nebulae is by measuring the brightness of variable stars which are called cepheids. Without going into details, it suffices to say that by observing the type of variability of these stars it is possible to discover their absolute brightness (luminosity), and that if such stars have the same period of variation their absolute brightness is also equal; their observed magnitudes may be different, since their distances from the earth may differ. According to Euclidean geometry the brightness of the observed star diminishes as the square of its distance from the earth grows, since the energy E of emitted light is distributed, at a distance r from its source, over the surface of a sphere of radius r, so that the quantity of energy $\frac{E}{4\pi r^2}$ falls onto a unit area of the surface.

Hence the conclusion: if two stars of the same absolute brightness have observed brightness I_1 and I_2, then

$$I_1 : I_2 = \frac{1}{r_1^2} : \frac{1}{r_2^2},$$

where r_1 and r_2 denote the distances of the stars from the earth. In practice the distance r_2 is already known from another source, since the second star is nearer the earth, and r_1, the distance of the further star, may be determined from the above proportion.

In Lobatchevskian geometry the area of a sphere is greater than $4\pi r^2$, so that the observed brightness of the star diminishes more quickly as r increases than in Euclidean geometry; it will therefore take the value given by observational evidence for a smaller distance r_1, i. e. the figures obtained according to Euclid will be too large. The degree to which they are too large depends on the value of the constant k, and for more remote spiral nebulae the differences would be greater.

PROBLEM. *Find the length of an arc of an equidistant.*

Let us imagine an equidistant whose points are distant from the straight line m by a constant segment c, and let the projection of the arc K_1L_1 on m have length p

(Fig. 110). Let us inscribe in the arc K_1L_1 of the equidistant a polygonal line with equal sides, one of which is CE. From the ends of the segment CE and from its centre D we drop perpendiculars to m and get the quadrangle $ABCD$ with three right angles. To this we apply the

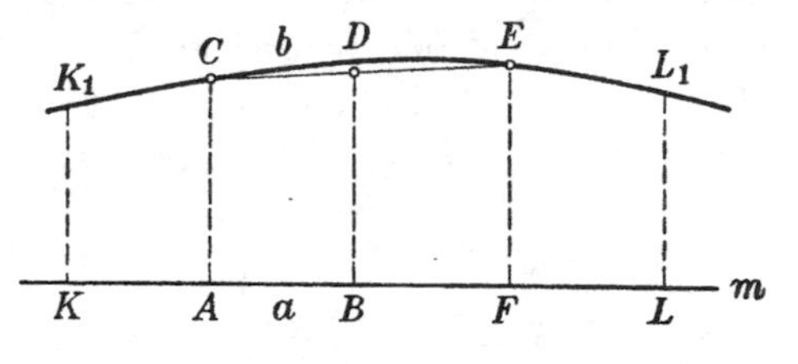

FIG. 110

results of the problem on p. 159. Write $AB = a$, $CD = b$. We had

$$\frac{\tanh a}{\tanh b} = \sin\alpha \quad \text{and} \quad \sin\alpha = \frac{1}{\cosh c},$$

therefore

$$\frac{\tanh b}{\tanh a} = \cosh c.$$

Let us replace the hyperbolic tangent by the ratio of the hyperbolic sine to the hyperbolic cosine:

$$\frac{\sinh b}{\sinh a} \cdot \frac{\cosh a}{\cosh b} = \cosh c.$$

If in this formula b and a tend to zero, then $\cosh a$ and $\cosh b$ tend to unity. The ratio $\dfrac{\sinh b}{\sinh a}$ may be written as

$$b \cdot \frac{\sinh b}{b} : a \cdot \frac{\sinh a}{a}.$$

It follows from the remark made at the end of § 19 that the ratio $\frac{\sinh b}{b}$ tends to 1 as b tends to zero, and so does $\frac{\sinh a}{a}$. Finally, therefore, if b and a are sufficiently close to zero the left-hand side and therefore also the right-hand side of our formula, i. e. $\cosh c$, approaches $\frac{b}{a}$.

The ratio of the length of the inscribed polygonal line to its projection p on m is exactly $\frac{b}{a}$, so that on increasing the number of sides of the polygonal line or, in other words, allowing b and a to tend to zero, this ratio tends to the limit $\cosh c$. The limit of the length of the polygonal line will be referred to as l, the length of the arc of the equidistant. It is given by the formula

$$l = p \cosh c,$$

which solves the problem.

In the subjects discussed in this section, as in the analogous topics of Euclidean geometry, an essential rôle is played by the theory of limits. To determine the length of an arc we inscribed a polygonal line in it, added its sides, and then queried whether this sum would tend to a definite limit as the number of sides tended to infinity. Our arguments were of a sporadic nature—we appealed a number of times to geometrical intuition (we did so, for instance, as we were discussing the area of a sphere: quadrangles on spheres were treated as if they were plane figures) and also to the theorems from § 19 assumed without proof. A careful analysis of this method and its establishment on general principles is a part of integral calculus, where it is given in a precise, and at the same time simple form. To this field of science belong in particular questions of calculating lengths of lines, areas and volumes. In Euclidean geometry, however, the idea of limits may be dispensed with in two important cases, namely, it is possible to develop without referring to the concept of limit—elementarily, so to say—the theory of areas of polygons and the theory of volumes of prisms.

In non-Euclidean geometry the development of the

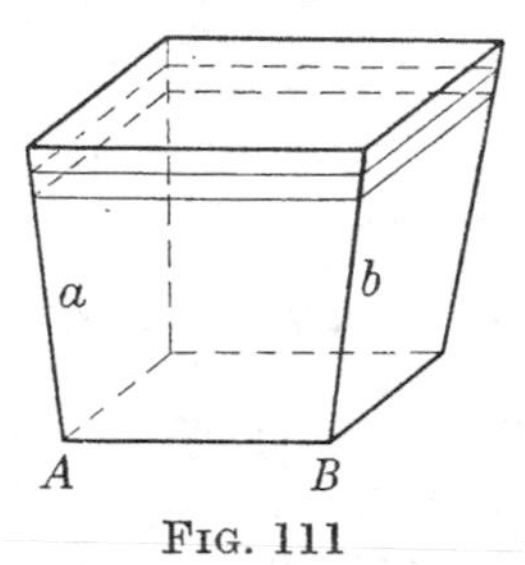

FIG. 111

theorems on the areas of polygons may also proceed in an elementary manner: *the area of a polygon is simply the product of its defect and the constant* k^2. On the other hand there is no type of polyhedron in non-Euclidean geometry whose volume can be computed in a simple, elementary way. Certainly we may construct a polyhedron (Fig. 111) whose base is, say, a square, and whose vertical edges are equal and perpendicular to the base, but it does not possess the important property of the simple prism, that the area of the section made by a plane cutting off equal segments on the vertical sides should be equal to the area of the base, since these edges, being perpendicular to the same straight line, are divergent and go away from each other.

Therefore, in order to determine the volume of a polyhedron, it must be cut up into "slices" and the volumes of the slices summed up. This is an intricate task involving integrating and the result of the calculations is rather complex.

As with prisms, so with other solids. Apart from the sphere only the volume of a right circular cone may be expressed easily. It is

$$V = \pi k^2 (l\cos\alpha - h), \tag{22}$$

where l is the side (generator) of the cone, h its altitude, α the angle between the side and the altitude.

This formula was discovered by Lobatchevsky, who devoted a large part of his entire writings (in fact, nearly half of his *Pangeometria*) to the calculation of areas, surfaces and volumes. At the same time Lobatchevsky proved by means of his non-Euclidean geometry a number of formulae of integral calculus, manifesting a great analytical skill and a thorough familiarity with current foreign mathematical writing. He attributed great significance to these ideas since he considered them a verification of the correctness of the science which he was establishing. "Thus all the arguments", he wrote in his voluminous dissertation *Applications of non-Euclidean geometry to cartain integrals*, "lead to consistent results and prove the legitimacy of the principles of non-Euclidean geometry" ([1]).

He hoped too that by proving non-Euclidean geometry to be useful in integral calculus he might win adherents among mathematicians. He did not succeed to convince his contemporaries, but to our minds the scientific and philosophical significance of his concepts is without doubt, and demands no support from applications which are in any case of less value.

§ 27. Spherical trigonometry

First let us recall some facts of elementary geometry. The term "trigonometry" originates from the Greek

---

([1]) Lobatchevsky used the term *Voobražaemaya geometriya* which means something like "imaginable geometry". In his last work he declared that "it is better to call this geometry "pangeometry", since this term implies a more general geometry, of which the ordinary one is a special case".

and means measurement of triangles. Plane trigonometry deals with triangles—that is, figures formed by three points and three lines connecting them—and spherical trigonometry deals with figures formed by three half-lines with the same origin and three angles connecting (one says usually: between) them. We assume here that our half-lines do not lie in the same plane.

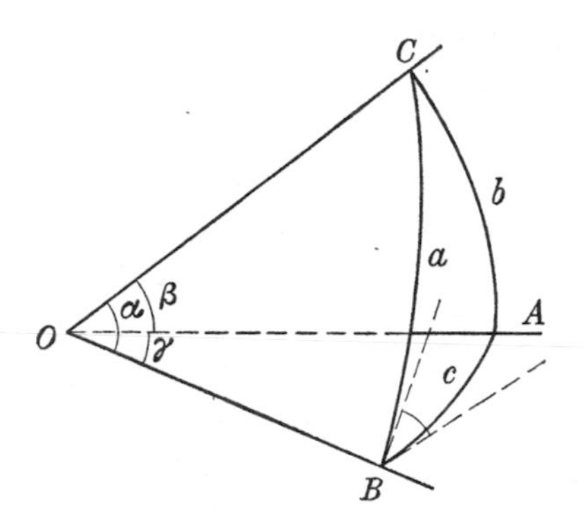

FIG. 112

Such a figure is called a *trihedron*, its *edges* are the half-lines and the angles between the edges are the face angles of *the trihedron*. The planes of these angles form dihedral angles, or dihedra for short, between them: these are the dihedra with edges OA, OB, OC in Fig. 112. Thus in each trihedron we have six elements: three face angles $\alpha = \measuredangle BOC$, $\beta = \measuredangle AOC$, $\gamma = \measuredangle BOA$ and three dihedra, which we shall denote by A, B, C. The angles $\alpha, \beta, \gamma, A, B, C$ are less than $180°$.

Let us draw a sphere of radius r and centre O. Its surface cuts each of the faces of the trihedron along an arc (AB, BC, CA in the figure). All these arcs together define a spherical triangle ABC on the surface of the sphere. The sides of this triangle are the arcs $AB = c$, $BC = a$, $CA = b$; the angles of the triangle are formed by these arcs and are said to be equal to the angles between the tangents to the arcs. The tangents are perpendicular to the radii of the sphere, so the angles between them are equal to the corresponding dihedral angles; for example, the angle of the spherical triangle at the vertex B is equal to the dihedron with edge OB, i. e. to B.

The spherical triangle therefore has sides a, b, c and angles A, B, C. The spherical triangle and the trihedron are connected with one another: the angles of the former are equal to the dihedra of the latter, and there is a relationship as well between the sides of the former and the face angles of the latter. It may be expressed in the simplest way if we assume that α, β, γ are radian-measures

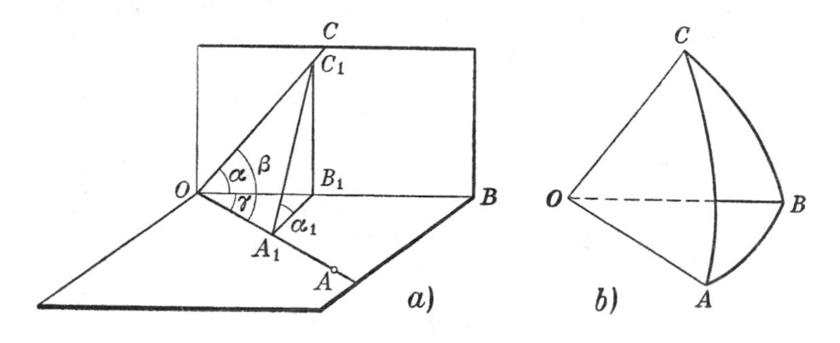

FIG. 113

of the face angles BOC, AOC, AOB. The radian-measure of a right angle is $\frac{1}{2}\pi$ and the length of the corresponding arc of circle is $\frac{1}{2}\pi l$, where l is the radius r in Euclidean, and $k \sinh \frac{r}{k}$ in non-Euclidean geometry. To an angle of α radians there corresponds an arc of length αl, so that in our case we have

$$a = \alpha l, \quad b = \beta l, \quad c = \gamma l.$$

The theory of trihedra may be treated as the theory of triangles on a sphere, whence originates the term spherical trigonometry. Certain relationships exist between the sides and the angles of such triangles. We will now deduce them.

As in plane trigonometry and in non-Euclidean trigonometry, the most important task is to find the equations connecting the sides and the angles of a right-angled triangle. Let, say, the angle B be right. Then the planes OBA and OBC are perpendicular to one another. This is shown in Fig. 113a, where the first plane is horizontal and the other vertical;

$$\sphericalangle BOA = \gamma, \qquad \sphericalangle AOC = \beta, \qquad \sphericalangle COB = \alpha.$$

From C_1 on OC we drop the perpendiculars C_1B_1 to OB and C_1A_1 to OA. A_1B_1 is the projection of A_1C_1 onto the horizontal plane. By the absolute theorem on three perpendiculars we have

$$B_1A_1 \perp OA_1.$$

The lines A_1C_1 and A_1B_1 lie on the faces of the dihedron with edge OA, and are perpendicular to OA. Therefore they contain an angle equal to the dihedral angle, i. e.

$$\sphericalangle C_1A_1B_1 = A.$$

From now on we proceed differently in Euclidean and non-Euclidean geometry. We shall give them parallel treatment.

First let us apply three times the fundamental theorem on right-angled triangles ((11), p. 143):

Euclidean geometry	Non-Euclidean geometry

First to the triangle $A_1B_1C_1$:

Euclidean geometry	Non-Euclidean geometry
$\sin A = \dfrac{B_1C_1}{A_1C_1}$,	$\sin A = \dfrac{\sinh \dfrac{B_1C_1}{k}}{\sinh \dfrac{A_1C_1}{k}}$,

second to the triangle OB_1C_1:

$$B_1C_1 = OC_1 \sin\alpha, \qquad \sinh\frac{B_1C_1}{k} = \sinh\frac{OC_1}{k}\sin\alpha,$$

third to the triangle OA_1C_1:

$$A_1C_1 = OC_1 \sin\beta, \qquad \sinh\frac{A_1C_1}{k} = \sinh\frac{OC_1}{k}\sin\beta.$$

We substitute the latter values in the first formula. In both cases we get

$$\sin A = \frac{\sin\alpha}{\sin\beta}. \tag{23}$$

Next we apply three times the theorem on right-angled triangles ((10), p. 143):

First to the triangle OB_1C_1:

$$OB_1 = OC_1 \cos\alpha, \qquad \tanh\frac{OB_1}{k} = \tanh\frac{OC_1}{k}\cos\alpha,$$

second to the triangle OB_1A_1:

$$OA_1 = OB_1 \cos\gamma, \qquad \tanh\frac{OA_1}{k} = \tanh\frac{OB_1}{k}\cos\gamma,$$

third to the triangle OA_1C_1:

$$OA_1 = OC_1 \cos\beta, \qquad \tanh\frac{OA_1}{k} = \tanh\frac{OC_1}{k}\cos\beta.$$

Substituting the first and third formulae in the second we get in both cases, after some abbreviation

$$\cos\beta = \cos\alpha\cos\gamma. \tag{24}$$

The above calculation and Fig. 113 assume that angles α and γ are acute. If this were not so some slight modifications (signs!) would have to be made, but the final result would be the same. The reader may verify this.

It is striking that the relationships between the angles A, α, β, γ have the same form in Euclidean as in non-Euclidean geometry. This might have been forecast, since very close to the point O Euclidean geometry is certainly in force and the angles A, α, β, γ are exactly the same in that neighbourhood as in the figure $OA_1 B_1 C_1$.

Since the formulae of spherical trigonometry are the same in both geometries it follows that their legitimacy does not depend upon whether the axiom of Euclid is true or not. Briefly:

Spherical trigonometry is independent of the axiom of Euclid, and therefore belongs to absolute geometry.

This property gives us a perfect example of an extensive geometrical theory independent of the axiom of Euclid.

One may replace the trihedra in the argument by the corresponding spherical triangles and state: both geometries give the same relationships between the sides and the angles of spherical triangles. On these relationships can be founded the study of geometrical figures on the sphere, i. e. the spherical geometry. We have used the same locution as in the section on the horosphere.

Spherical geometry is the same in Lobatchevskian as in Euclidean geometry.

Let us turn again to formulae (23) and (24) and apply them to the spherical triangle (Fig. 113b), whose angles at the vertices A, B and C are equal to the dihedra A, 90°, C, and whose sides are $a = \alpha l$, $b = \beta l$, $c = \gamma l$ (cf. formulae of p. 179).

$$\sin A = \frac{\sin \dfrac{a}{l}}{\sin \dfrac{b}{l}}, \qquad \cos \frac{b}{l} = \cos \frac{a}{l} \cos \frac{c}{l},$$

$$\sin \frac{a}{l} = \sin \frac{b}{l} \sin A, \tag{25}$$

$$\cos \frac{b}{l} = \cos \frac{a}{l} \cos \frac{c}{l}. \tag{26}$$

In these formulae $\cos\frac{a}{l}$ means the cosine of the angle whose radian-measure is $\frac{a}{l}$; and so on.

From these formulae several consequences follow by simple working-out; for example, by replacing $\sin A$ in the first formula by $\sqrt{1-\cos^2 A}$, and $\sin\frac{a}{l}$ by $\sqrt{1-\cos^2\frac{a}{l}}$, and evaluating $\cos\frac{a}{l}$ from the second, we obtain the relationship

$$\tan\frac{c}{l} = \tan\frac{b}{l}\cos A, \tag{27}$$

which is valid for acute and also for obtuse values of the angle A. The reader may verify this.

The spherical triangle ABC is a right-angled triangle (Fig. 114); the angle B is right, $BC = a$, $BA = c$ are its sides and $AC = b$ is its hypotenuse. The angle A lies opposite a and next to c.

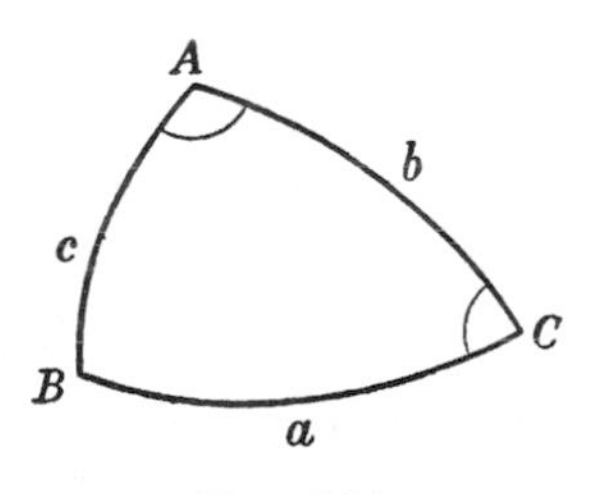

FIG. 114

The ratio $\frac{c}{l}$ may be looked on as the side measured in terms of the unit l, and then the relationship (27) formulated as follows:

The tangent of the short side (measured in terms of the unit l) *is equal to the product of the tangent of the hypotenuse* (measured in terms of l) *and the cosine of the angle adjacent to this side.*

Formula (25) gives:

The sine of the short side (measured in terms of l) *is equal to the product of the sine of the hypotenuse* (measured in terms of l) *and the sine of the angle opposite this side.*

These are fundamental theorems about right-angled spherical triangles. They are the basis not only for the solution of right-angled spherical triangles but also for that of any spherical triangle, since a triangle may be divided into two right-angled ones. The gist of these theorems covers the whole of spherical geometry.

We do not propose to develop the theorems of this science nor to discuss its extremely wide applications in the fields of astronomy, geodesy or crystallography. We shall be concerned only with the absorbing fact which has certainly not escaped the notice of the reader: that the fundamental theorems of spherical trigonometry are the very twins of those of hyperbolic trigonometry.

Let us compare them (see pp. 180 and 181).

Non-Euclidean trigonometry	Spherical trigonometry	
$\tanh\dfrac{c}{k} = \tanh\dfrac{b}{k}\cos\alpha,$	$\tan\dfrac{c}{l} = \tan\dfrac{b}{l}\cos A,$	(28)
$\sinh\dfrac{a}{k} = \sinh\dfrac{b}{k}\sin\alpha,$	$\sin\dfrac{a}{l} = \sin\dfrac{b}{l}\sin A.$	

In both cases α and A denote that angle of the triangle which lies opposite the side a.

The formulae on the left will result from those on the right if we replace the constant l by the constant k and the trigonometrical functions of sides by the hyperbolic functions of sides. We shall understand the state of things better if we take a look at the relationships between the functions themselves (see § 19):

$\cosh^2 y - \sinh^2 y = 1,$	$\cos^2 y + \sin^2 y = 1,$
$\cosh(y+z)$	$\cos(y+z)$ (29)
$= \cosh y \cosh z + \sinh y \sinh z.$	$= \cos y \cos z - \sin y \sin z.$

It is obvious that if we write, in the right-hand formulae (29), hyperbolic cosines for cosines and for sines the ratio of a hyperbolic sine to $\sqrt{-1} = i$, we shall be left with the left-hand formulae. The division by i is necessary for the sake of the signs in the lower formulae.

The same substitution in the right-hand formulae (28) applied to the functions of sides a, b, c will give the parallel left-hand formulae. To be sure, the factor i appears (since we substituted on both sides of the equality the ratios of the hyperbolic sines of the sides to i for the sines of the sides), but it may be cancelled out.

Summarizing: the equations of spherical trigonometry will be transformed into those of Lobatchevskian trigonometry if we substitute the constant l for the constant k and at the same time the hyperbolic cosines of sides for the ordinary cosines and the hyperbolic sines divided by $\sqrt{-1}$ for the sines.

Thus, if a certain equation of spherical trigonometry contains the cosines of sides $c_1, c_2, c_3, \ldots$ and the sines of sides $s_1, s_2, s_3, \ldots$, then the corresponding equations of Lobatchevskian trigonometry will be obtained by the transformation described above.

The converse is obviously also valid. We have, for example, proved for all triangles on the Lobatchevskian plane the cosine theorem (formula (19), p. 167):

$$\cosh a = \cosh c \cosh b - \sinh b \sinh c \cos \alpha$$

(with a suitable choice of unit).

Replacing cosh by cos and sinh by $\sqrt{-1}$. sin we immediately obtain the cosine theorem of spherical trigonometry

$$\cos a = \cos b \cos c + \sin a \sin c \cos A.$$

The spherical and Lobatchevskian trigonometries are thus in some way geared together: the equations of one linking $c_1, c_2, \ldots, s_1, s_2, \ldots$ become the equations of the other by the simple transformation $c \to c$, $s \to si$, that is, by a simple change of variables. Therefore, if we wish to examine these equations algebraically, e. g. to eliminate an unknown from them, they will appear equivalent in both theories. The geometrical interpretation of the roots of the equations may involve differences which do not occur in an algebraic investigation—owing to the fact that the roots should lead to real geometric quantities.

The connection between the equations of spherical and those of Lobatchevskian geometry could be put in a very suggestive way. It would, however, be understood only by the reader familiar with the following formulae, known as *Euler's formulae*:

$$\cos a = \frac{e^{ia} + e^{-ia}}{2}, \qquad \sin a = \frac{e^{ia} - e^{-ia}}{2i}.$$

Applying these to $\cos\frac{a}{l}$ and $\sin\frac{a}{l}$ we get

$$\cos\frac{a}{l} = \frac{e^{i\frac{a}{l}} + e^{-i\frac{a}{l}}}{2}, \qquad \sin\frac{a}{l} = \frac{e^{i\frac{a}{l}} - e^{-i\frac{a}{l}}}{2i}.$$

Let us write ki for l:

$$\cos\frac{a}{ki} = \frac{e^{\frac{a}{k}} + e^{-\frac{a}{k}}}{2} = \cosh\frac{a}{k},$$

$$\sin\frac{a}{l} = \frac{e^{\frac{a}{k}} - e^{-\frac{a}{k}}}{2i} = \frac{\sinh\frac{a}{k}}{i}.$$

The upshot of this is that if we put ki instead of l in the formulae of spherical trigonometry we will get hyperbolic cosines for ordinary cosines and hyperbolic sines divided by i for sines—in other words, the same change that leads from the formulae of spherical trigonometry to those of Lobatchevskian trigonometry.

In Euclidean geometry l is the radius of the sphere. When we replace l by ki we introduce, so to speak, an imaginary radius of the sphere. Therefore we may express the result as follows:

Lobatchevskian trigonometry coincides with trigonometry on a sphere of imaginary radius.

Of course, this sentence should be taken as nothing more than an expressive description of the relationship between the equations of Lobatchevskian and of spherical trigonometry. This connection is one of the most interesting facts of non-Euclidean geometry.

§ 28. Analytical geometry

An exposition of non-Euclidean geometry would be incomplete without some mention of the elements of analytical geometry. This was the opinion of Lobatchevsky, who discussed in his *Pangeometria*, among other things, the equations of the circle and straight line.

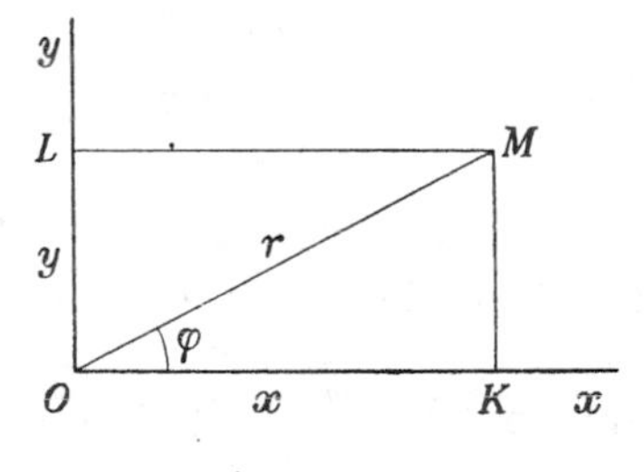

FIG. 115

This subject causes no difficulty, but at the very beginning we notice a characteristic difference from Euclidean geometry. In the latter the Cartesian coordinates of a point M may be defined in two ways:

Either we may draw from M a perpendicular MK (Fig. 115) to the axis Ox of the Cartesian system Oxy and denote by x the length of OK and by y the length of KM, or else we draw from M perpendiculars MK and ML to both axes and denote by x, y the lengths of OK and OL respectively.

It is unimportant which we do, since $MK = OL$.

This is not the case in non-Euclidean geometry, where the segment OL, being perpendicular to both OK and LM, is shorter than MK.

We shall adopt the second method (Lobatchevsky used the first), since both axes of coordinates will then have equal rights. *The coordinates of point M are the lengths (supplied with suitable signs) of the projections of the vector OM onto both axes.*

It follows from this definition that each point of the plane has a definite pair of coordinates x, y, but the converse does not hold true. In fact, we have seen in the problem discussed on p. 162 that the perpendicular at the point K to Ox will cut the perpendicular to Oy at L if and only if

$$\tanh^2 x + \tanh^2 y < 1 .$$

(Throughout this section we will omit the denominators k.)

Therefore *the numbers x, y are the coordinates of a definite point on the plane if and only if they satisfy the above inequality.*

Let us denote by r the distance of M from the centre of the system, and by φ the angle between the vector OM and the positive direction of the axis Ox.

From the triangles MOK and MOL it follows from the theorem on right-angled triangles (p. 143)

$$\begin{aligned} \tanh x &= \tanh r \cos\varphi, \\ \tanh y &= \tanh r \cos(90^\circ - \varphi) \\ &= \tanh r \sin\varphi . \end{aligned} \tag{30}$$

It is easy to show that these relationships hold for any point M on the plane.

Let us square formulae (30) and add them. We get

$$\tanh^2 r = \tanh^2 x + \tanh^2 y. \tag{31}$$

Formula (31) might be called the formula for the distance of a point from the origin.

It is now easy to solve the following problems:

PROBLEM. *Find the equation of a circle with centre at the origin and radius R.*

We have only to state that the distance of the point (x, y) from the origin is R, therefore by (31):

$$\tanh^2 x + \tanh^2 y = \tanh^2 R, \tag{32}$$

which is the wanted equation.

PROBLEM. *Find the equation of a circle with centre at an arbitrary point $A(a, b)$ and radius R.*

We must express the fact that the segment connecting the points $A(a, b)$ and $M(x, y)$ has length R (Fig. 116). Write

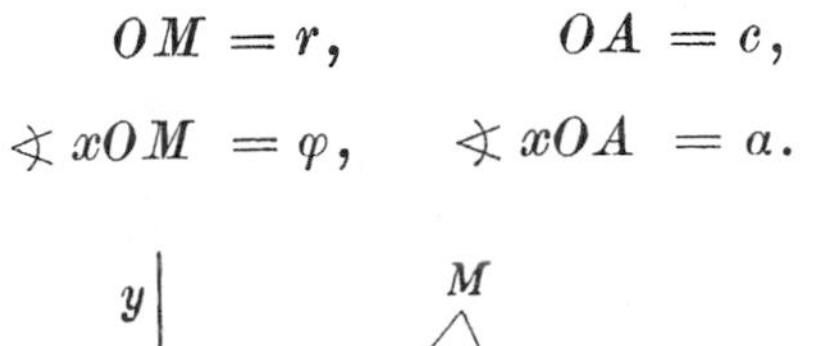

$$OM = r, \qquad OA = c,$$
$$\sphericalangle xOM = \varphi, \qquad \sphericalangle xOA = \alpha.$$

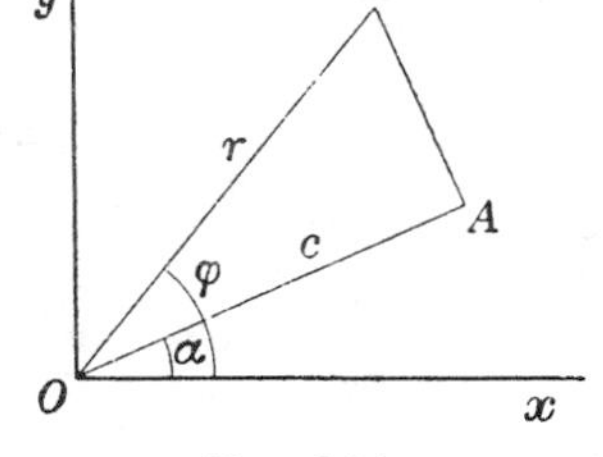

FIG. 116

We find the distance AM by using the cosine theorem (formula (19), p. 167) to the triangle OAM:

$$\cosh AM = \cosh r \cosh c - \sinh r \sinh c \cos(\varphi - \alpha)$$
$$= \cosh r \cosh c [1 - \tanh r \tanh c \cos(\varphi - \alpha)].$$

The second term in square brackets is equal, by the formula

$$\cos(\varphi - \alpha) = \cos\varphi\cos\alpha + \sin\varphi\sin\alpha$$

to the expression:

$$\tanh r\cos\varphi\tanh c\cos\alpha + \tanh r\sin\varphi\tanh c\sin\alpha.$$

Now by formulae (30) we have

$$\tanh r\cos\varphi = \tanh x, \qquad \tanh c\cos\alpha = \tanh a.$$

The first term of this expression is equal to $\tanh x\tanh a$, the second similarly to $\tanh y\tanh b$, whence the whole expression is

$$\tanh x\tanh a + \tanh y\tanh b.$$

In order to determine the distance AM we must now consider the factor $\cosh r$. We shall proceed as in ordinary trigonometry: dividing both sides of the equation

$$\cosh^2 r - \sinh^2 r = 1$$

by $\cosh^2 r$, we get

$$1 - \tanh^2 r = \frac{1}{\cosh^2 r}.$$

Hence, and by (31), we get

$$\cosh^2 r = \frac{1}{1 - (\tanh^2 x + \tanh^2 y)}.$$

Similarly

$$\cosh^2 c = \frac{1}{1 - (\tanh^2 a + \tanh^2 b)}.$$

Finally, the formula for $\cosh AM$ will assume the form

$$\cosh^2 AM = \frac{(1 - \tanh x\tanh a - \tanh y\tanh b)^2}{(1 - \tanh^2 x - \tanh^2 y)(1 - \tanh^2 a - \tanh^2 b)}. \tag{33}$$

This is the formula for the distance between two points with given coordinates (a, b) and (x, y).

We then immediately obtain the equation of the wanted circle

$$\cosh^2 R = \frac{(1-\tanh x \tanh a - \tanh y \tanh b)^2}{(1-\tanh^2 x - \tanh^2 y)(1-\tanh^2 a - \tanh^2 b)}. \quad (34)$$

PROBLEM. *Find the equation of a straight line, given its distance $c = OA$ from the centre of the system and the angle between OA and the x-axis* (Fig. 117).

The triangle MOA is a right-angled triangle; its hypotenuse is $OM = r$, its side is $OA = c$, and the angle

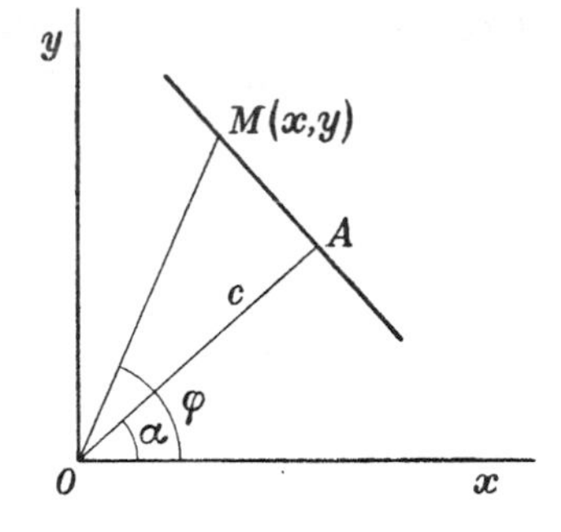

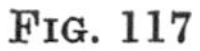
FIG. 117

between these sides is the difference $(\varphi - a)$. Therefore, by the theorem of p. 143 we get

$$\tanh OA = \tanh OM \cos(\varphi - a).$$

And conversely, if this relation is satisfied the triangle OAM is right-angled.

As in the last problem,

$$\tanh c = \tanh r \cos\varphi \cos a + \tanh r \sin\varphi \sin a,$$

but

$$\tanh r \cos\varphi = \tanh x, \qquad \tanh r \sin\varphi = \tanh y,$$

whence

$$\tanh c = \tanh x \cos a + \tanh y \sin a. \quad (35)$$

This equation is satisfied by the points of a straight line, and conversely a point satisfying this equation lies on it. Thus the equation of a straight line is of the form of (35).

As we see, all the equations we have obtained are quite intricate: for instance, the equation of a circle with centre O and radius R, if we bear in mind the definition of the hyperbolic tangent (p. 130), will be

$$\left(\frac{e^x - e^{-x}}{e^x + e^{-x}}\right)^2 + \left(\frac{e^y - e^{-y}}{e^y + e^{-y}}\right)^2 = \text{const}.$$

Nevertheless, the situation is not so frightening as might appear at first sight. The coordinate x appears in both the equation of a circle and that of a straight line in the term $\tanh x$. The same applies to the coordinate y.

This suggests the introduction of new variables

$$X = \tanh x, \qquad Y = \tanh y. \tag{36}$$

Using these the equation of a straight line will assume the form

$$X \cos a + Y \sin a = \text{const}, \tag{37}$$

and the equation of a circle with centre at the origin, the form

$$X^2 + Y^2 = \text{const}. \tag{38}$$

Finally, the equation of any circle is

$$(1 - AX - BY)^2 = \cosh^2 R (1 - X^2 - Y^2)(1 - A^2 - B^2). \tag{39}$$

We have put here $A = \tanh a$, $B = \tanh b$.

These last equations do not look so odd: the equation of the straight line is of the first degree, that of the circle of the second degree with respect to X, Y. We have discovered an amazing fact: that the Cartesian coordinates which reigned over Euclidean analytical geometry lose their monopoly of simplicity in Lobatchevskian. On the contrary—even the circle with centre O has a complex equation when expressed in these coordinates.

We are not sorry to forsake them preferring the variables X and Y. What shall we call them? Quite simply, the coordinates of a point. For what is most essential in the concept of the coordinates of a point if not the fact that we can find the point when we know them? For instance, if we know the latitude and longitude of a place we are able to find it without difficulty. The coordinates X and Y certainly possess this property. We shall limit ourselves to their positive values.

$X = \tanh x$ is an increasing function of x assuming every value between 0 and 1, and these once only, as x ranges from 0 to ∞. The same is true for $Y = \tanh y$. Therefore, given X and Y it is possible uniquely to determine x and y and then to find the point M as in Fig. 115. This is possible (see p. 163) if

$$\tanh^2 x + \tanh^2 y < 1,$$

i. e., if

$$X^2 + Y^2 < 1. \tag{40}$$

If the numbers X and Y satisfy the above inequality they will uniquelly determine a point. They are called *Beltrami's coordinates*, since he was the first to realize their convenience. Lobatchevsky also used them on occasion but without laying emphasis on them.

Thus the points of the plane have been uniquely associated with the pairs of numbers (X, Y) which satisfy the inequality $X^2 + Y^2 < 1$. The equation of a straight line is linear with respect to the variables X, Y. Further, the equation of a circle with centre O is

$$X^2 + Y^2 = \text{const},$$

and the distance between points having the Beltrami coordinates (X, Y) and (A, B) is expressed by the formula

$$\cosh^2 R = \frac{(1 - AX - BY)^2}{(1 - X^2 - Y^2)(1 - A^2 - B^2)}. \tag{41}$$

For instance, the distance between points whose Beltrami coordinates are $\frac{8}{9}$, $\frac{4}{9}$ and $\frac{2}{3}$, $\frac{2}{3}$ may be evaluated from

$$\cosh^2 R = \frac{\left(1-\frac{8}{9}\cdot\frac{2}{3}-\frac{4}{9}\cdot\frac{2}{3}\right)^2}{\left(1-\frac{4}{9}-\frac{4}{9}\right)\left(1-\frac{64}{81}-\frac{16}{81}\right)} = \frac{\left(\frac{1}{9}\right)^2}{\frac{1}{9}\cdot\frac{1}{81}} = 9,$$

whence

$$\cosh R = 3, \qquad \text{and from the tables} \qquad R = 1{\cdot}758.$$

The further development of analytical geometry on the Lobatchevskian plane would require the discussion of a number of questions, e. g. how to recognize whether two straight lines with given equations will cut, and how to find the formula for the angle between them; how to find the equation of the perpendicular to a given line, or the distance of a given point from a given line. All this lies beyond the scope of this book, for we have confined ourselves to discussing the most characteristic principles of analytical geometry on the Lobatchevskian plane.

§ 29. The Klein model

In this section we shall discuss a question of extreme importance. We have taken a look at a number of topics arising in Lobatchevskian geometry and have learnt the solutions of many problems, the fruit of genius and creative phantasy, and we have touched on the staggering concepts of modern physics; in all this we have silently passed over the fundamental question which has been under discussion for centuries. Is it possible to prove the axiom of Euclid by showing that its negation leads to a contradiction? Admittedly this has not so far been done, but that does not mean that it never will be. Man is always looking for ever newer discoveries of which never dreamt his ancestors. The human mind goes on from triumph to triumph; perhaps one day it will succeed in proving that Lobatchevskian geometry contains interval inconsistencies?

Lobatchevsky was convinced that this was not so, but since he could find no evidence in empirical reality that the assumptions of non-Euclidean geometry were satisfied there, he comforted himself with the fact that the applications of this geometry lead to correct results in integral calculus. Of course, this argument was not enough to settle the question, although to some extent it did support Lobatchevsky's ideas. Finally, he attempted in his memoir *Voobražaemaya geometriya* to give a rigorous proof of the consistency of the principles of his geometry, arguing that "it is impossible to obtain from them false results in any respect".

The new exposition made use of facts known from elementary trigonometry. We all know that from the fundamental theorems dealing with right-angled triangles (the side is equal to the product of the hypotenuse and the sine of the opposite angle or the cosine of the adjacent angle) are derived the theorems for all triangles; from this arises the possibility of proving geometrical theorems by means of "computation". For instance, it is possible to prove that the three altitudes of a triangle intersect at one point by evaluating the segments cut by two altitudes on the third and showing that the first of these segments is equal to the second.

The same is also true in non-Euclidean and in spherical trigonometry. From both theorems on right-angled triangles (§ 20) we deduce the theorems for (oblique-angled) triangles (e. g. cosine theorem) and various other geometrical theorems. It is, for instance, possible to show in this way that the medians of a triangle intersect in one point. The theorems for right-angled triangles contain in latent form, as the seed contains the future plant, our entire knowledge of triangles.

Lobatchevsky, for instance, shows that if positive numbers a, b, c and positive angles α, β, γ (less than π)

satisfy the cosine theorem (of non-Euclidean geometry) then the sum of the angles is less than 180°. Thus the theorem on the sum of the angles of a triangle is a logical consequence of the equations which express the cosine theorem.

Observing this, Lobatchevsky conceived the original idea of constructing non-Euclidean geometry on purely analytical, or calculative foundations: namely, he set down the trigonometrical theorems for right-angled triangles as basic assumptions, i. e. as axioms, and then deduced from them the theorems of geometry—for example, the theorem on the sum of the angles of a triangle.

Geometry so formulated loses its character of a synthetic exposition and becomes a framework of formulae deduced from each other by means of algebraic arguments. In such a series of deductions it is impossible to find a contradiction unless the initial equations were contradictory. A theory obtained in this way may not be true—in the sense of not agreeing with empirical reality—but will certainly contain no inconsistencies within itself. Lobatchevsky proved that the initial equations were not contradictory by appealing to spherical trigonometry (as in § 28). This point of his argument is not essential, for the same can be shown without employing spherical trigonometry.

Lobatchevsky's idea was penetrating and perfectly correct—later proofs of the non-contradictoriness of non-Euclidean geometry followed a similar pattern—but its presentation, however, in *Voobražaemaya geometriya* was too sketchy and not fully convincing. It is not easy to see what geometrical facts he has assumed together with his trigonometrical equations, and it is not entirely apparent what his argument is based on. Not until many years after Lobatchevsky's death did the human mind attain clarity in this subject and work out a method to prove that a certain set of theorems will never lead to inconsistent corollaries.

This method follows the direction indicated by Lobatchevsky: the investigation of a certain framework of concepts with easily provable lack of self-contradiction, and which is closely bound up with the theory under consideration.

In order to explain this better let us recall that in § 28 a pair of numbers X, Y, subject to the inequality

$$X^2 + Y^2 < 1,$$

was uniquely associated with every point on the Lobatchevskian plane, and that conversely a definite point on this plane corresponds to every such pair of numbers.

Let us now imagine the Euclidean plane α with a system of Cartesian coordinates on it. Points on this plane with coordinates X, Y fill the interior of a circle k of radius 1. The Lobatchevskian plane is mapped into this area (this is another mapping than that discussed in § 10, since the circle in § 10 was a part of the Lobatchevskian plane while the present one is a part of the Euclidean plane). Let us examine this mapping more closely.

The equation (37) of a straight line on the Lobatchevskian plane is linear with respect to X, Y, so it is the equation of a straight line in the plane α. It follows that, in our map, straight lines of the Lobatchevskian plane become chords of the circle. Points with coordinates (X, Y) and (A, B) on our map correspond to the points on the Lobatchevskian plane the distance between which (R) is given by the formula

$$\cosh^2 R = \frac{(1 - AX - BY)^2}{(1 - X^2 - Y^2)(1 - A^2 - B^2)}. \tag{42}$$

It is easy to ascertain whether two triangles on our map correspond to congruent or non-congruent triangles on the Lobatchevskian plane—it suffices to calculate by the above formula the lengths of the sides of both triangles (in the Lobatchevskian plane) and verify that they are equal. By the same argument we may also discover whether two angles on our map represent equal or different angles on the Lobatchevskian plane, since the question of whether angles are equal may always be reduced to an examination of triangles.

Various problems of the Lobatchevskian plane may be solved on the "map". Let us take two points on the map with coordinates (0,0) and ($\frac{1}{2}$, 0), for example. They correspond to two points on the Lobatchevskian plane—the ends of a certain segment. To the problem of bisecting this segment there corresponds on the map the problem of finding such a point (X, O) whose "distances" from the points (0, 0) and ($\frac{1}{2}$, 0), as calculated by the formula (41), would be equal. The former distance will be given by the formula

$$\cosh^2 R_1 = \frac{(1-X.0-Y.0)^2}{(1-X^2-0)(1-0-0)} = \frac{1}{1-X^2},$$

and the latter by

$$\cosh^2 R_2 = \frac{(1-X.\frac{1}{2}-Y.0)^2}{(1-X^2-0)(1-\frac{1}{4}-0)} = \frac{(1-\frac{1}{2}X)^2}{\frac{3}{4}(1-X^2)}.$$

As these distances are to be equal we shall write

$$\frac{(1-\frac{1}{2}X)^2}{\frac{3}{4}} = 1,$$

whence

$$X = 2 \pm \sqrt{3}.$$

The sign + would give $X > 1$, but since X is the coordinate of a point within the circle of radius 1, we have

$$X = 2 - \sqrt{3}.$$

We have found that point on the "map" which corresponds to the centre of the segment in question on the Lobatchevskian plane.

In our working appeared only the coordinates of points of the map and we made our calculations according to the normal procedure of analytical geometry with the only difference that we took the "distance" between the points from formula (42) and not from the formula $\sqrt{(X-A)^2+(Y-B)^2}$.

Similarly, to various problems of the Lobatchevskian plane there correspond definite problems on the "map". Drawing a straight line through two points becomes drawing a chord through two points; determining two segments which would be equal to each other becomes the finding of two segments on the map whose lengths, as given by formula (42), would be equal. If we wished to construct an angle equal to a given one we should have to take equal segments on its sides and examine the resulting triangle. All these problems lead, on the map, to definite computations resting on formula (42).

Let us now turn the whole matter upside-down. Let us forget that any Lobatchevskian plane exists and any straight lines or points at any given distances on it. We consider for a while only the circle k of radius 1 on the Euclidean plane—its chords and the points lying within it. We shall call the chords straight lines, and by the distance between two points, i. e. the length of a segment, we shall mean the positive number R which satisfies equation (42); triangles will be considered as equal if they have equal sides. In this scheme to problems and proofs of geometry correspond frameworks of consecutive formulae. These formulae cannot lead to any contradiction, since all our working is based on the properties of the Euclidean plane, on adopted definitions and on algebraic transformations. The term "length" for the number R does not matter. Of course, we are taking for granted that Euclidean geometry contains no inconsistencies.

Very nice, the reader may say, I am prepared to admit that the arguments concerning the circle k and the problems connected with it cannot lead to a contradiction, and that the spelling of the term adopted for a certain concept is not important, but what of that?

The answer is, in fact, no less than that the properties of this map exactly coincide with the properties of the Lobatchevskian plane. Two points may be connected by a straight line. A given segment may be cut off on a straight line on either side of a given point. The "length" of the sum of two consecutive segments on a straight line, evaluated by the formula (41), is equal to the sum

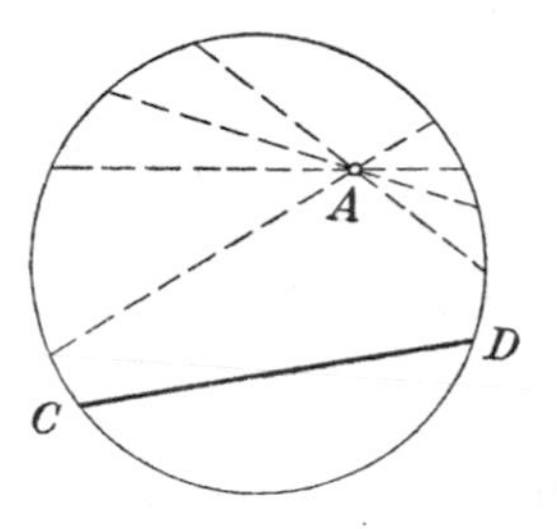

FIG. 118

of the "lengths" of those segments. Triangles with equal sides have equal angles. Through a point A not lying on a straight line CD there passes an infinite number of straight lines which do not cut this straight line; these are, in Fig. 118, the chords passing through A and not cutting CD. Because our "map" geometry cannot lead to a contradiction, no more can Lobatchevskian geometry.

Of course, we have not proved all the theorems stated above; we have not shown, for instance, that we are in fact able to cut off on each straight line a segment of any length. A fuller and more precise exposition should fill these gaps. It was our intention to give at least an overall picture of how the fundamental problem of the noncontradictoriness of Lobatchevskian geometry has been settled.

Let us characterize the method, the first germ of which may be traced back to Lobatchevsky himself. In order to show that a certain set of theorems contains no inconsistency we construct an object whose suitably-defined elements satisfy the theorems of this set, and then assert that the theorems of our set are consistent since they hold in the object. This object we call a model of the set of theorems in question. The circle k together with associated definitions is known as the *Klein model* of non-Euclidean geometry. In modern science various kinds of models like this play very important rôle.

APPENDICES

1. We shall base the proof of the theorem that *plane angles of a dihedron are equal* on the theorem that *figures symmetrical with respect to the centre of symmetry O are congruent*, the simplest case of which, dealing with the equality of symmetrical segments (Fig. 119), is the corollary to the equality of vertical angles. The equality of symmetrical triangles (therefore also of symmetrical angles) follows from the preceding theorem by the equality of corresponding sides.

Let ABC and DEF (Fig. 120) be plane angles of the dihedron, and O be the centre of the segment BE. We construct the angle A_1EC_1 symmetrical to the angle ABC about O, and therefore equal to it

$$\sphericalangle A_1EC_1 = \sphericalangle ABC.$$

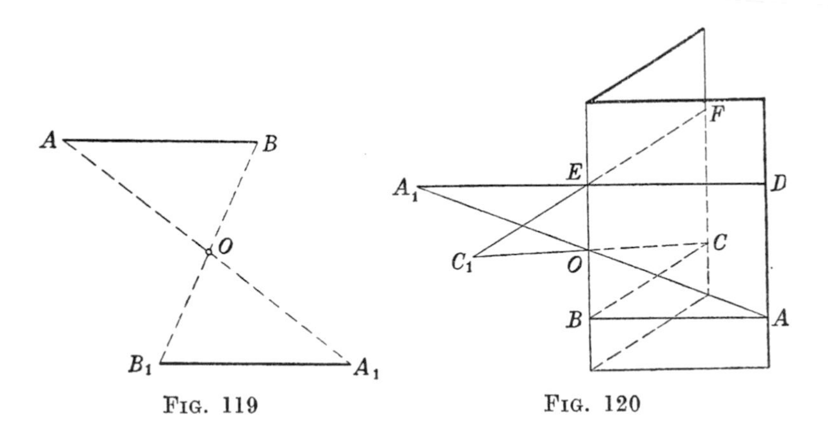

Fig. 119

Fig. 120

On the other hand $\sphericalangle A_1EO$ is symmetrical to the right angle ABO, whence

$$A_1E \perp EO.$$

$DE \perp EO$ by assumption, and A_1E and ED lie in one plane; therefore DE and EA_1 lie on one straight line. Similarly FEC_1 is a straight line. The angles DEF and A_1EC_1 are vertical angles. Therefore, and from the first equality, it follows that $\sphericalangle ABC = \sphericalangle DEF$. Q. E. D.

2. The functions $\cos t$ and $\sin t$ refer to the circle of radius 1. The point M with coordinates $x = \cos t, y > \sin t$ (Fig. 121) lies on the circle described by the equation

$$x^2 + y^2 = 1.$$

As t ranges from 0 to $\frac{1}{2}\pi$, the point M travels along the quadrant from A to B. The number t may be interpreted either geometrically as the radian-measure of the angle AOM or as twice the area of the segment AOM.

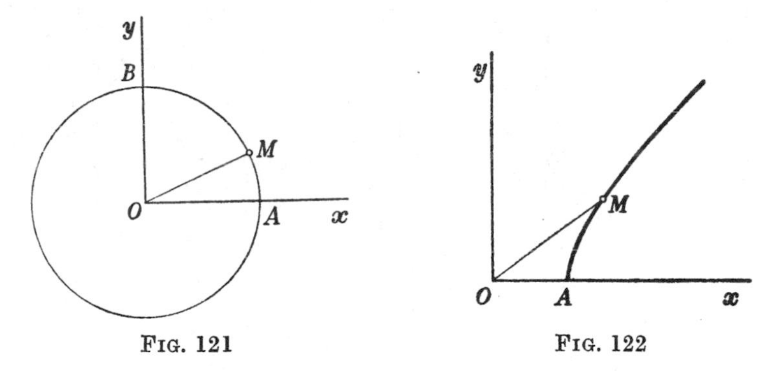

Fig. 121 Fig. 122

Similarly, the point M with coordinates $x = \cosh t, y = \sinh t$ lies on the hyperbola described by the equation

$$x^2 - y^2 = 1.$$

As t ranges from 0 to ∞, the point M travels along the quarter-hyperbola from A to ∞ (Fig. 122). The number t cannot be interpreted as the angle AOM, but it may be proved that it is equal to twice the area of the hyperbolic segment AOM... For this reason we refer to the functions $\cosh t$ and $\sinh t$ as *hyperbolic functions*.

Table of values of hyperbolic functions

u	$\sinh u$	$\cosh u$	$\tanh u$	$\coth u$
0·00	0	1	0	∞
0·02	0·0200	1·0002	0·0200	50·007
0·04	0·0400	1·0008	0·0400	25·013
0·06	0·0600	1·0018	0·0599	16·687
0·08	0·0801	1·0032	0·0798	12·527
0·10	0·1002	1·0050	0·0997	10·033
0·12	0·1203	1·0072	0·1194	8·373
0·14	0·1405	1·0098	0·1391	7·189
0·16	0·1607	1·0128	0·1586	6·303
0·18	0·1810	1·0162	0·1781	5·615
0·20	0·2013	1·0201	0·1974	5·066
0·22	0·2218	1·0243	0·2165	4·619
0·24	0·2423	1·0289	0·2355	4·246
0·26	0·2629	1·0340	0·2543	3·932
0·28	0·2837	1·0395	0·2729	3·664
0·30	0·3045	1·0453	0·2913	3·433
0·32	0·3255	1·0516	0·3095	3·231
0·34	0·3466	1·0584	0·3275	3·054
0·36	0·3678	1·0655	0·3452	2·897
0·38	0·3892	1·0731	0·3627	2·757
0·40	0·4107 (1)	1·0811	0·3799	2·632
0·42	0·4325	1·0895	0·3969	2·519
0·44	0·4543	1·0984	0·4136	2·417
0·46	0·4764	1·1077	0·4301	2·325
0·48	0·4986	1·1174	0·4462	2·241
0·50	0·5211	1·1276	0·4621	2·164
0·52	0·5437	1·1383	0·4777	2·093
0·54	0·5666	1·1494	0·4930	2·028
0·56	0·5897	1·1609	0·5080	1·969
0·58	0·6131	1·1730	0·5227	1·913
0·60	0·6366	1·1855	0·5370	1·862
0·62	0·6605	1·1984	0·5511	1·814
0·64	0·6846	1·2119	0·5649	1·770

(1) Final figure in heavy type indicates that following figure is 5.

Table of values of hyperbolic functions (*continued*)

u	$\sinh u$	$\cosh u$	$\tanh u$	$\coth u$
0·66	0·7090	1·2258	0·5784	1·729
0·68	0·7336	1·2402	0·5915	1·691
0·70	0·7586	1·2552	0·6044	1·655
0·72	0·7838	1·2706	0·6169	1·621
0·74	0·8094	1·2865	0·6291	1·589
0·76	0·8353	1·3030	0·6411	1·560
0·78	0·8615	1·3199	0·6527	1·532
0·80	0·8881	1·3374	0·6640	1·506
0·82	0·9150	1·3555	0·6751	1·481
0·84	0·9423	1·3740	0·6858	1·458
0·86	0·9700	1·3932	0·6963	1·436
0·88	0·9981	1·4128	0·7064	1·416
0·90	1·0265	1·4331	0·7163	1·396
0·92	1·0554	1·4539	0·7259	1·378
0·94	1·0847	1·4753	0·7352	1·360
0·96	1·1144	1·4973	0·7443	1·344
0·98	1·1446	1·5199	0·7531	1·328
1·0	1·1752	1·5431	0·7616	1·313
1·1	1·3356	1·6685	0·8005	1·249
1·2	1·5095	1·8107	0·8336	1·199
1·3	1·6984	1·9709	0·8517	1·160
1·4	1·9043	2·1509	0·8853	1·129
1·5	2·1293	2·3524	0·9054	1·105
1·6	2·3756	2·5775	0·9217	1·085
1·7	2·6456	2·8283	0·9354	1·069
1·8	2·9422	3·1075	0·9468	1·056
1·9	3·2682	3·4177	0·9562	1·046
2·0	3·6269	3.7622	0·9640	1·037
2·1	4·0219	4·1443	0·9704	1·030
2·2	4·4571	4·5679	0·9757	1·025
2·3	4·9370	5·0372	0·9801	1·020
2·4	5·4662	5·5569	0·9837	1·017
2·5	6·0502	6·1323	0·9866	1·014
2·6	6·6947	6·7690	0·9890	1·011
2·7	7·4063	7·4735	0·9910	1·009
2·8	8·1919	8·2527	0·9926	1·007
2·9	9·0596	9·1146	0·9940	1·006
3·0	10·0179	10·0677	0·9950	1·005

Table of values of hyperbolic functions *(continued)*

u	$\sinh u$	$\cosh u$	$\tanh u$	$\coth u$
3·2	12·246	12·287	0·9967	1·003
3·4	14·965	14·999	0·9978	1·002
3·6	18·285	18·313	0·9985	1·001
3·8	22·339	22·362	0·9990	1·001
4·0	27·290	27·308	0·9993	1·001
4·2	33·336	33·351	0·9995	1·0004
4·4	40·719	40·732	0·9997	1·0003
4·6	49·737	49·747	0·9998	1·0002
4·8	60·751	60·759	0·99986	1·0001
5·0	74·203	74·210	0·99991	1·0001
5·2	90·633	90·639	0·99994	1·0001
5·4	110·701	110·705	0·99996	
5·6	135·211	135·215	0·99997	
5·8	165·148	165·151	0·99998	
6·0	201·713	201·716	0·99999	

INDEX